VECTOR PARTICLE PHYSICS

T.N. Lockyer
Author/Publisher

TNL PRESS

PUBLISHER: TNL PRESS
1611 Fallen Leaf Lane
Los Altos, CA 94024

First edition 1992

9 8 7 6 5 4 3 2 1

Digit on the right indicates the number of this printing.

Library of Congress Catalog Card Number: 91-90733

ISBN: 0-9631546-0-5
ISBN: 0-9631546-1-3 (pbk)

PRINTER: **BookCrafters**, Chelsea, MI 48118

FOR

Virginia

and

Robert, Richard and Timothy

Patricia, Carol and Bonnie

and

Cynthia, Melissa, Christine, Aubrey and Melanie

Brady, Brett, Derek and Brian

INTRODUCTION

When you can measure what you are speaking about and express it in numbers, you know something about it. And when you cannot express it in numbers, your knowledge is of a meagre and unsatisfactory kind. It may be the beginning of knowledge, but you have scarcely, in your thought, advanced to the stage of a science.
Lord Kelvin (1824-1907)

We use numbers to separate the scientific from the specious theories of physics. Accordingly, this monogram demonstrates that these vector particle models give numerically correct predictions, for all related fundamental physical constants, to the same precision those constants are known.

By applying Lord Kelvin's standards, you will be able to verify that these vector particle models are correct by definition. You will need a computer or a scientific calculator for review of these vector models.

Vector Particle Physics presents, for the first time, mathematically consistent models for the structures of energy and matter. The models reveal an internal structure for the photon, electron, positron, electron type neutrino pair, muon type neutrino pair, and the composite proton, neutron and muon.

These proposed vector particle models are the first serious challenge to QCD. QCD has been called the "standard model" for particle composition. If you are familiar with the standard model, then you know that QCD has fatal flaws, for example, the failure to make precision predictions for mass and magnetic moments. It is known that *all* particles that have mass *also* exhibit a magnetic moment. Lacking the ability to accurately predict mass and magnetic moment, is, of course, extremely embarrassing to QCD. In my view, any particle model that fails to accurately predict mass and magnetic moment is a waste of time and money. Another thing that concerns me, about the "standard model", is its ad hoc nature. For example, there are no structures or coherence to accurately scale. Consequently, the standard model *WILL NEVER BE ABLE TO MAKE PRECISION NUMERICAL PREDICTIONS.* QCD can only be a provisional model that must (sooner or later) be replaced with a correct model.

What has been lacking, in the current theories of physics, is a logical approach to the modeling process. Accordingly, we must start at a layer below that of the standard model by giving a structure to the basic particles. Everyone agrees that energy forms matter. It follows, logically then, that we must first model energy itself. With a model for energy, we can then see clearly if it is possible to structurally justify postulating the basic quarks of the standard model. As will be demonstrated, the quarks are *NOT* possible basic constructions, only three pair of basic leptons are possible. The arguments are outlined in these summarized CHAPTERS.

CHAPTER 1 introduces the photoproduction of the electron positron pair. The creation of matter by photons is the primary thrust of this book. Any particle with mass must be composed of energy (photons). Deducing the structural model for energy is the secret to deducing the structures of all subatomic particles. The chapter lists the fixed ratios between the fundamental physical constants and gives some brief arguments to show that both QCD and QED are mathematically *INCONSISTENT* theories.

CHAPTER 2 provides a vector model for the photon (energy) as the orthogonal coupling of two conjugate (zero energy) (E,B) and (B,E) resonances. The photon is shown to be a dynamic Poynting vector that alternates from rest to the square root of two times (c) velocity as it translates. The velocity of light (c) becomes an RMS (effective)value. The photon model has two types of constructions that make possible the modeling of *BOTH* attraction and repulsion. Attraction and repulsion, by the exchange of photon momentum, is demonstrated in Chapter 5.

CHAPTER 3 develops structures for the leptons, using the photon's vectors. Only three pair of particles are possible constructions, because we exhaust all possible ways to combine the photon's vectors. The resulting natural boundary conditions, which limits particle types, produces a controlled model that mimics nature. (In contrast, the standard model lacks such structural constraints. Quarks, having no structure, are ad hoc and out of control.)

CHAPTER 4 shows that one can justify and predict *ALL* of the electron's fundamental physical constants from the orthogonal geometry of a spinning cube, giving credibility to the vector model concept. The electron models as a cube with an edge length equal to the electron's rationalized Compton wavelength.

CHAPTER 5 discusses the specious claims for having detected the "ghostly" neutrinos, and discusses the new revelations about the neutrino's characteristics, as provided by the vector models. The neutrino would seem to be a much more important particle, in the scheme of things, than previously suspected. The neutrino model gives access to the nested cubes of hyperspace, from the neutrino model's ability to shrink to sub space size. Access to subspace dimensions, by the neutrino, allows the composite particles to form.

CHAPTER 6 provides the theory of vector combination rules between the electron and the neutrino. The neutrino model shrinks to subspace size by absorbing energy and forms in the vortex of the spinning electron. The concepts of charge conjugation and scaling in the electron's spinning vortex are introduced. Electrical potential energy between conjugating particles is discussed.

CHAPTER 7 provides the detailed internal structures for the proton and neutron vector models. The models give the mass ratio of the proton to within 3.24 ppm, and the neutron's mass ratio to within 0.4 ppm of the CODATA recommendations. Moreover, the model predicts the known beta decay products of the neutron, when it becomes a proton. The electron and electron type neutrino decay products come from the unformed pairs at the first two structural layers of the neutron model. These several results came out of the models as a natural consequence of the model coherence, they were not put into the models. Getting results that were not put into these vector models are indications that these vector models are paralleling nature.

CHAPTER 8 The last scaled adjustment to the proton model adds about 28 percent to the model's mass ratio. That (last model adjustment) is a proton core particle that can predict the proton's magnetic moment to within 33 parts per million of the CODATA value. (There has been no previous method to relate proton mass to proton magnetic moment.) The exponentially increasing proton model truncates after adding the last 28 percent of its mass as a core particle. It would seem that the proton core particle is the smallest (hence most massive) construction possible in nature. If we postulate heavier basic particles, we may be writing checks that nature can't cash. (For example, we can write down a velocity of 3 $x10^9$ meters per second, but nature can't comply. A similar natural boundary may prevent individual particles from being heavier than the truncated proton core particle.)

CHAPTER 9 develops the complex structure for the neutron that allows the neutron to have a magnetic moment, negative helicity, and, at the same time, exhibit its zero external charge.

CHAPTER 10 shows the structure of the muon, from the vector combination rules, as an electron or positron that has been collapsed by the muon type neutrinos. The lack of charge conjugation, by a muon type neutrino's vectors in concert with the electron, is shown. The resulting absence of electrostatic attraction, to hold the the muon's structural parts together, explains why the muon comes apart. (decays)

CHAPTER 11 discusses the mysterious fine structure constant and presents several ways to calculate the dimensionless ratio as being of geometric origin. The electron model's magnetic moment acts like a two turn electromagnet. The electromagnet's geometry allows the calculation of the fine structure constant, from the Ampere turns per meter.

APPENDIX A provides an over-view of the MKSA (SI) system of units used exclusively in this monogram, and gives a listing of selected fundamental physical constants.

APPENDIX B provides a computer program for computing the proton and neutron models. Included in the program is an algorithm for computing consistent sets of constants, from a small subset of exactly known constants.

The inescapable message, taken from these models, is that energy forms only primary leptons and it is the neutrinos that allow leptons to form composite particles at greater mass (energies). The resulting precision and coherence of these models is taken as proof of the concepts.

WILL VECTOR PARTICLE PHYSICS PREVAIL?

I am painfully aware that it may be a long time before the specious standard models are replaced with these more correct vector particle models. The only particle theories that find acceptance, it seems, are those that require building bigger and bigger accelerators. The politics, serving that agenda, does not always serve scientific correctness. It is hoped that the engines of greed, fed by big science, will gradually be replaced by the engines of reason.

Being retired and a free-lance writer, removes me from any political networks. I have no political debts to pay, nor colleague reputations to preserve. Having the freedom to call things, as they logically occurred, has allowed these models to develop in a natural, orderly and unfettered fashion. I hope you are pleased with the results.

If you are pleased with these vector models, I would ask your help to establish them into the larger scientific community.

Thomas N. Lockyer
December 3, 1991

VARIATIONS ON A THEME

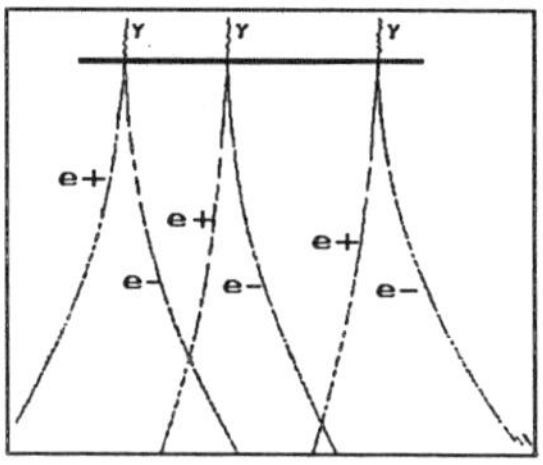

SETTING THE STAGE:

The chapter heading graphics shows nature in the act of creating matter from gamma ray photons (γ) in the presence of a target. In this case the target is in the form of a lead plate. The mass of the target is required in order to create the necessary "photon fireball". Here the gamma ray photons (γ) have created three electron positron pairs. The particles are shown as separated from their mates, as by a magnetic field normal to the picture's surface. The unknown detail, about this pair production process, is what mechanism energy uses to always create the same particle types. The electron and positron are always created in pairs and at exactly the same mass (energy) values. Electrons and positrons are created in the same absolute size anywhere in the universe.

At this interface between energy and matter, during matter formation, nature (mysteriously) sets the (exact) values for the fundamental physical constants. The best estimates, for those exact constants, are collected by the National Institute for Science and Technology (NIST) formerly (NBS), and promulgated by the (international) Committee on Data for Science and Technology (CODATA). Values for selected constants are given in the APPENDIX A, Table 1.0.

Every electron in the universe has exactly the same mass (m_e) spin angular momentum ($\frac{1}{2}\hbar$) electrical charge (e) and magnetic moment (μ_e). Further, the electron is the source of Planck's constant (h) as a consequence of the electron being a Planck sized chunk of mass energy. The reader can verify that ($h = [m_e c^2] [\lambda_e / c]$) from APPENDIX A values. The electron is viewed as a harmonic oscillator, in black body radiation. Planck's constant, derived from black body radiation, is the manifestation of the electron's Planck size.

The characteristics of the electron allow calculating the values for all related constants, exactly, using a number of precise mathematical equations. (*Figure 1.0*) By algebra one can extend the expressions in (*1.0*) to any single unknown. (See APPENDIX A values)

$$\epsilon_0 = \mu_0^{-1} c^{-2} \qquad e = \sqrt{\frac{2\alpha h}{\mu_0 c}}$$

$$\lambda_e = \frac{h}{m_e c} \qquad \lambda_e = \frac{\alpha^2}{2 R_\infty}$$

NOTE: R_∞ is the Rydberg $= 1.09737315 \times 10^7 \text{ m}^{-1}$

FIGURE 1.0 Relations among selected physical constants.

The energy (E) of each of the individual photons is given by ($E = hf$). Since the frequency (f) can take on any value, the photon energy is a virtual continuum. Just what mechanism nature uses to always create the minimum energy electron positron pairs, from out of the photon's high energy fireball, remains one of the great unsolved mystery of physics.

Particles with a mass greater than the mass of an electron are composites and (ultimately) decay into their fermion constituents. This being the case, and nature being restricted to creating "quantum sized chunks" of basic particles, composite particles (like the proton or neutron) must, of necessity, be composed of minimum energy basic particles. This is a viable approach, because an electron (lepton) comes out of the neutron to form a proton, and, conversely, an electron (lepton) goes into a proton (K shell electron capture for example) to form a neutron. Leptons go into hadrons, and come out of hadrons. It seems a reasonable conclusion then, that leptons are inside of the hadrons in some form.

Logically, from the demonstrated photoproduction of particles, leptons must be structured out of the structure of energy (the photon) itself.

If we were to postulate basic particles, without seeing how energy structured those particles, then that model would be doomed to only making approximate predictions for mass and magnetic moment. Those predictions would have to be based (entirely) on measured parameters. Unfortunately, the standard model's postulated quarks lack the required geometric structures. The standard model, to produce its (approximate) results, requires some 21 empirically derived parameters. These are: six quark masses, three lepton masses, W mass, Higgs mass, strong coupling constant, fine structure constant, Weinberg angle, 3CKM angles, 1CKM phase, ThetaQCD, LambdaQCD and Planck mass. Moreover, the model requires 36 different basic quarks with no proof that 36 quarks are possible constructions by energy.

It should also be mentioned that the quarks are given 1/3 or 2/3 of the fundamental charge. But, we can calculate the value of the fundamental charge from given fundamental constants. (*Figure 1.1*) If we are to believe in fractional charges, which constant in (*1.1*) would change? More importantly, why?

$$\alpha = 7.297353080929128 \times 10^{-3}$$
$$h = 6.626075470853031 \times 10^{-34}$$
$$\mu_0 = 4\pi \times 10^{-7}$$
$$c = 2.99792458 \times 10^{8}$$
$$e = \sqrt{\frac{2\alpha h}{\mu_0 c}} \qquad (\mathrm{A \cdot s})$$
$$e = 1.602177333892292 \times 10^{-19}$$

FIGURE 1.1 The indivisible fundamental charge from indivisible fundamental constants.

Quantum Electrodynamics (QED) treats the electron as a point particle, and the positron as an electron going backward in time. Point particles lead to mathematical difficulties from the resulting infinities. We will show that the electron has a structure, when modeled with electric (E) and magnetic (B) vectors. The positron becomes a conjugate structure which has a (B) and (E) where the electron has a (E) and (B) vector, respectively.

QED adjusts the coefficients of the power series of (α/π) by summing over Feynman diagrams of probability amplitudes. QED claims precise agreement with the experimentally measured electron magnetic moment. The coefficients obtained are assigned the following values: +1/2, -.328478444, +1.1763, -.8 , strange values to be sure.

Normal, expected values for a power series of the fine structure (α) divided by (π) would be +1/2, -1/3, +1/4, -1/5 etc. It can be shown that very good agreement with the published value for the electron's magnetic moment can be obtained using normal (non QED) coefficients. The reader can verify this as follows: (*Figure 1.2*)

$$(\alpha/\pi) = (Z) = 2.322819628 \times 10^{-3}$$

$$c = 2.99792458 \times 10^{8} \ (\mathrm{m \cdot s^{-1}})$$

$$e = 1.60217733 \times 10^{-9} \ (\mathrm{A \cdot s})$$

$$\lambda_e = 3.86159323 \times 10^{-13} \ (\mathrm{m})$$

$$\alpha_e = +\left(\left(\tfrac{1}{2}\right)(Z)\right) - \left(\left(\tfrac{1}{3}\right)\left((Z)^2\right)\right) + \left(\left(\tfrac{1}{4}\right)\left((Z)^3\right)\right) - \left(\left(\tfrac{1}{5}\right)\left((Z)^4\right)\right) + \left(\left(\tfrac{1}{6}\right)\left((Z)^5\right)\right)$$

$$\alpha_e = 1.159614444537 \times 10^{-3}$$

$$\mu_B = \tfrac{1}{2}(\lambda_e \cdot e \cdot c)$$

$$\mu_e = \mu_B(1+\alpha_e)$$

$$\mu_e = 9.284769731808 \times 10^{-24} \ \mathrm{J/T}$$

$$(9.2847701 \times 10^{-24} \ \mathrm{J/T})$$

(CODATA)

FIGURE 1.2 Normal (without QED) electron magnetic moment calculations to 42 ppb agreement.

The CODATA value for the electron's magnetic moment is listed in parenthesis. Note, in *FIGURE 1.2*, the 42 parts per billion agreement between the virgin (without QED) natural power series, and the experimental value. The 42 ppb is about the same precision with which the electron's magnetic moment may be measured.

It would appear that QED claims for precision are specious.

CUTTING TO THE CHASE:

The present theories of physics are wanting for lack of a structure for both energy and matter. In the next chapter we will develop a rudimentary model for the structure of energy, and from this we will deduce the structure for each subatomic particle that last longer than 2.2 microseconds. We will view all short lived particles as nature's mistakes which nature quickly corrects, and that such particles, lasting only a few microseconds, have no purpose in the scheme of the physical universe.

A MODEL FOR ENERGY

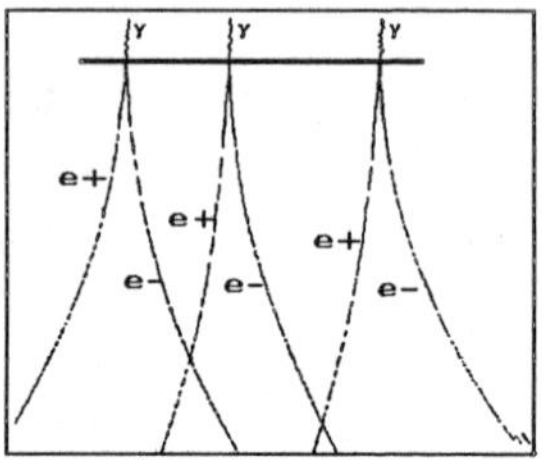

In this monogram, we will construct models for the long lived fermions. These are the electron, neutrinos, proton, neutron, muon, and their anti particles.

BACKGROUND:

Thomas Young (1773-1829) is credited with being the first to assign the term ENERGY to the product (mass times the velocity squared) and to set the work expended (the product of force times distance) proportional to the energy. Later, James Clerk Maxwell (1831-1879) identified the electromagnetic nature of energy. We have known since Rutherford (1871-1937) that matter and energy are interchangeable. What has been lacking is a good model for the structure of energy. Maxwell tried, shortly after developing the mathematics, to model electromagnetic energy. Unfortunately, Maxwell had a mistaken belief in a conducting "aether" making his energy model incorrect.

It can be demonstrated that energy (the photon) with an energy greater than twice the rest mass energy of the electron (or positron) can create an electron positron pair, in the presence of matter. The photon has to be turned back (Compton scattered) so as to form what we can term as a fireball. The fireball is necessary so that the photon's structure can latch to other photons and thus form particles. The chapter heading is a replica of a cloud chamber record of such a process. *Figure 2.0* shows a cloud chamber record of pair photoproduction.

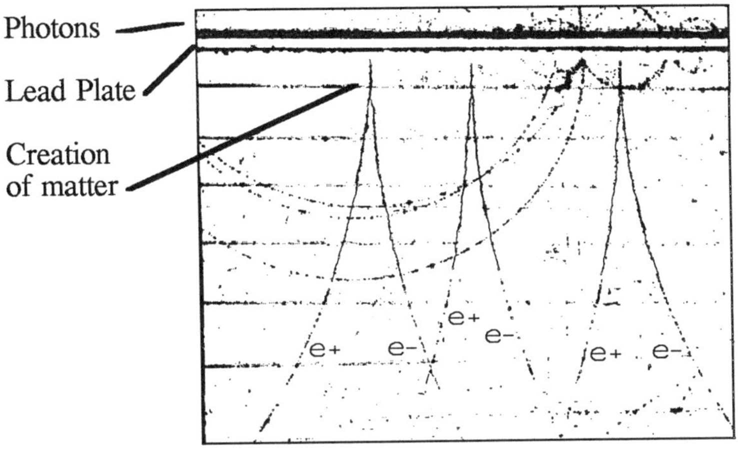

FIGURE 2.0 Typical cloud chamber record of electron positron pair production by X rays. (Courtesy of Lawrence Berkeley Laboratory)

We must visualize the structure of energy (the photon) so that we may see clearly how it is possible for the photon to crystallize into particles of matter.

STANDING ELECTROMAGNETIC WAVES:

Maxwell's well known equations describe electric and magnetic fields. A changing electric field (E) produces a magnetic field (B). This is the symmetrical counterpart of Faraday's law (i.e. a changing magnetic field produces an electric field). The E and B are thus referred to as being mutually dependent and inseparable, each owing its existence to the time rate of change of the other. The amplitudes of E and B, in an oscillating system can be expressed as:

E=Em cosine (Ø) B=Bm sine (Ø),

Where: Em=peak electric field in volts per meter,
Bm= peak magnetic field in Tesla
(Ø)=degrees from 0 to 360° of one cycle.
Em=cBm where c is the velocity of light.

A standing wave is shown schematically in *Figure 2.1* .

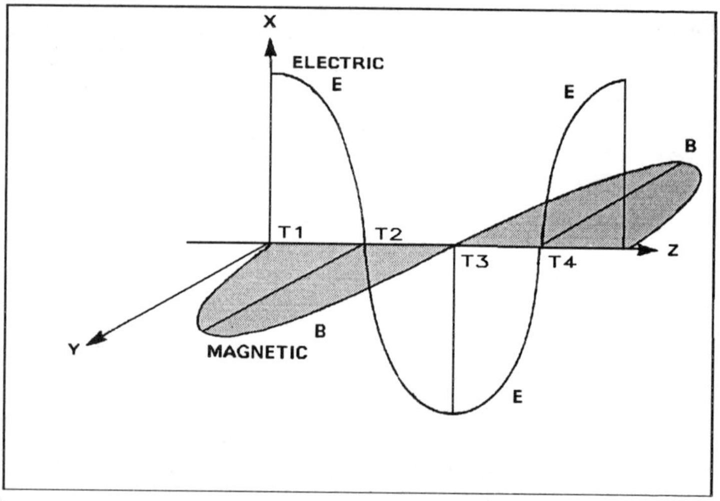

FIGURE 2.1 Standing electromagnetic wave.

The E and B of the standing wave are mutually dependent and inseparable. Each E and B amplitude, at any instant is proportional to the time rate of change of the other. For example, in *Figure 2.1*, on the Z axis, at time T1, the electric (E) is stationary at top dead center, (zero rate of change) and (B) is zero amplitude at that point. Conversely, at time T2, the electric (E) passes (at its maximum rate of change) from the negative quadrant to the positive quadrant, producing a maximum amplitude in the magnetic (B) at that point. On the other hand, at time T3, it is (B) at a maximum rate of change producing a maximum amplitude for (E). At time T4, the (B) reaches top dead center (zero rate of change) and (E) is zero amplitude.

ENERGY MODEL:

We have seen in *Figure 2.0* that energy (photons) structure matter (pairs). To model this process, we must deduce a model for the photon's structure. With a model for the photon's structure, we may then see clearly how particles form and how many are possible before exhausting all possible photon combinations. We will approach the modeling of the photon by visualizing what the photon sees from within its own relativistic frame of reference. *Figure 2.2* .

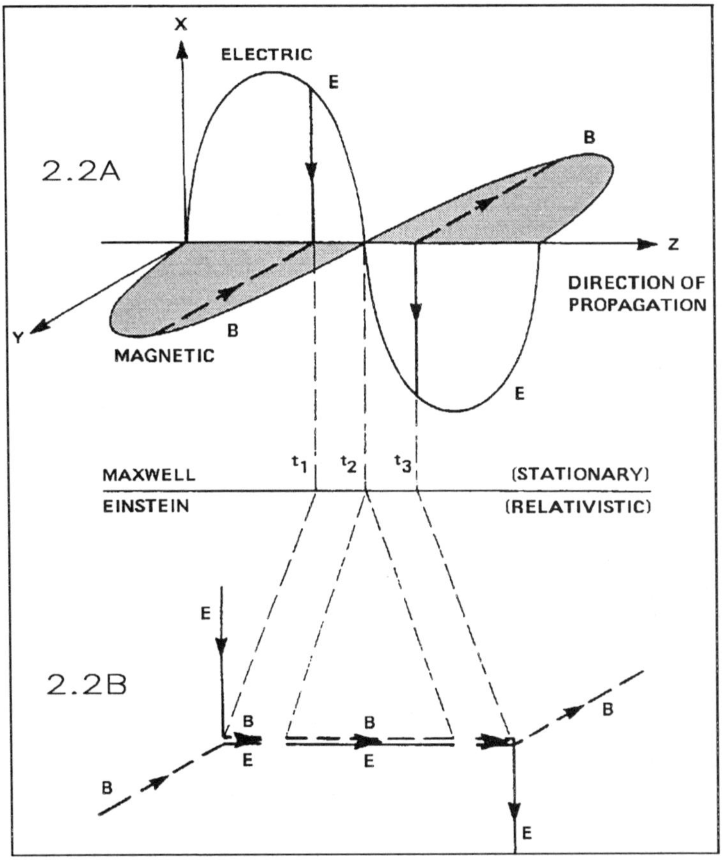

FIGURE 2.2 Traveling wave of electromagnetic energy.

Figure 2.2A shows a single cycle of a traveling wave of electromagnetic energy. An energy field is made up of a large number of photons, but the observation of a single photon is possible, and is also represented by *Figure 2.2A*. We immediately see two seemingly paradoxical things about the traveling wave, when viewed from our stationary frame of reference. Notice, for example, that

E and B seem to be *IN PHASE*. They should be 90 degrees out of phase to be mutually dependent and inseparable (as in *Figure 2.1*). Also notice that E and B both pass through zero at the same time. *THIS IS WRONG*, there is no ether, and energy cannot disappear without being replaced with matter.

Applying relativity leads to a vector model for the photon's structure. In *Figure 2.2B* we deduce the photon's structure from within its own relativistic frame of reference. The photon's deduced structure explains the apparent paradoxes in *Figure 2.2A*. First, the time coincidence, at time t_1 results from the coupling of two conjugate (E,B and B,E) resonances in the photon's structure! Lateral events are not affected by relativity, so we see the actual (E from one and B from the other) time coincidence in *Figure 2.2A*, at time t_1. The second paradox, E and B both simultaneously disappearing from our stationary frame of reference, at time t_2, is the result, of course, of the Lorentz Fitzgerald contraction at the extreme relativistic velocity. At time t_3, the lateral E, from one, and B, from the other conjugate resonances, have budded back into our stationary frame of reference view. Poynting (1852-1914) was the first to point out the vector properties of the rate of energy transport (proportional to the cross product of E X B). The photon (energy) model deduced in *Figure 2.2B* takes on the characteristics of a "dynamic, cyclic Poynting vector".

Note that the dynamic photon vector model is actually stationary twice each cycle as the lateral E,B vectors both reach top dead center, and are at maximum amplitude. Motion, then, in an absence of a conducting medium (an ether) is accomplished by the photon extending itself from one stationary point, in its defined space, to another, inch worm style. The photon's velocity, then, is effectively (c) but actual velocity goes from zero to the square root of two times (c). This photon model suggests the possibility of two types of photons. (1) the axial (Poynting) vector is pushed ahead of the structure. (2) the axial (Poynting) vector is trailing behind the structure. Consequently, we may now postulate positive and negative photon fields that model attraction or repulsion exchange forces, between particles. Like photons cause repulsion, and unlike photons cause attraction. Attraction has never before been modeled. A visual structure for the photon makes a theory possible.

SUMMARY:

We are able to model the photon's structure when we view the photon from the perspective of its relativistic frame of reference. The paradoxical sine, sine relationships between lateral E and B in the traveling wave is shown to be the result of the coupling of two conjugate (E,B) and (B,E) resonant pair together. Those conjugate resonances each (individually) have the required sine, cosine mutual embrace (E to B) and (B to E) and, at the same time, their lateral members exhibit the observed sine, sine (time coincidence) between the E of one and the B of the other.

In the next chapter we will use the photon's model (as a dynamic Poynting vector) to deduce the structure of the leptons (electron, electron type neutrino, muon type neutrino and their antiparticles).

THE PHOTON STRUCTURE OF LEPTONS

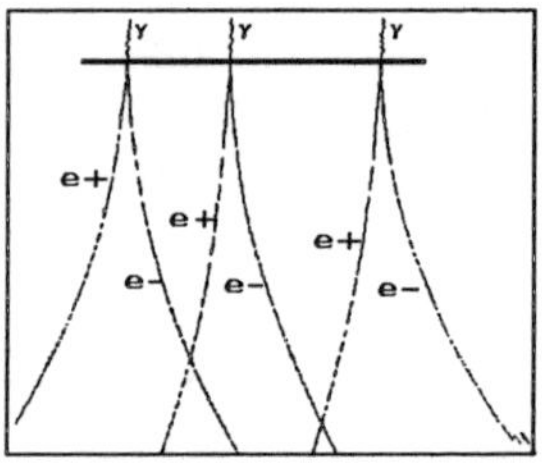

THE SHAPE OF PARTICLE MODELS:

To preserve the orthogonal arrangement that exists between E and B in nature, we will proceed to construct particle models in the form of vector cube frameworks. The belief that subatomic particles should be spherical may be aesthetically pleasing, but nature is under no obligation to agree with our preconceived notions. Accepting this, then, we can proceed to exhaust all possible ways to connect the photon to itself, and see where that approach will lead us.

We saw that the photon's vectors were thrusting at a maximum of ($\sqrt{2}$) times (c) so that the published value for (c) is actually a RMS or root mean square quantity. (APPENDIX A, Table 1.0). As a consequence, it can be shown, by the calculus of related rates, that nature will construct *ABSOLUTELY PERFECT* cubes from coupling of the photon's orthogonal vectors. This can be demonstrated by setting the Poynting vector length equal to (x) and E or B vector length equal to (y) around the perimeter of a cube face. The calculus of related rates indicates that the perimeter will stabilize at the instant that one edge lengthens while the other shortens at exactly (c). *Figure 3.0* .

The perimeter $P=2(x+y)$

Differentiating $\frac{dP}{dt}=2\left(\frac{dx}{dt}+\frac{dy}{dt}\right)$

Setting $\frac{dP}{dt}=0$ then $\frac{dx}{dt}=-\frac{dy}{dt}$

FIGURE 3.0 Calculus of related rates assures the cube face perimeter stablizes into a perfect square.

TYPES OF CONSTRUCTION:

I find that the use of colored pencils, for example *RED* as the Poynting edge, *BLUE* as the E edge and *GREEN* as the B edge allows rapid sketching to check the following construction details. The direction arrows can be added later.

The requirement for sharing the E and B vectors with a cube's four Poynting vector edges, is so that their thrust directions are continuous around the loop. In other words, an E and B must flow into and out of each Poynting vector edge. This necessary (natural) arrangement of vectors limits the possible constructions to just the three cube face types shown in *Figure 3.1* .

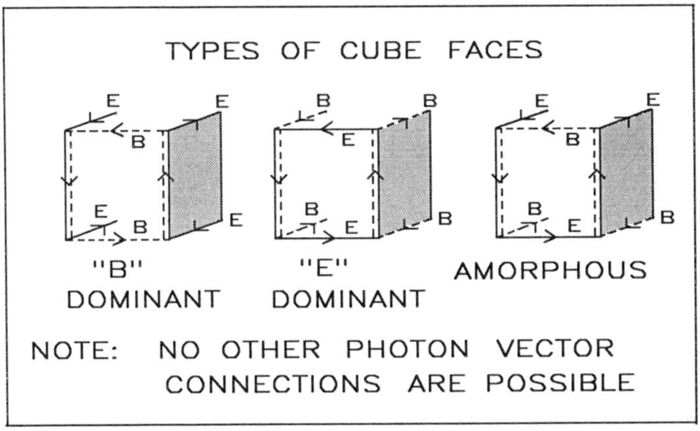

FIGURE 3.1 Natural boundary on cube face types.

The arrows show a counter clockwise direction, but this becomes a clockwise direction when viewed from behind. Spatial face location, on the cubes, then, gives twice as many possibilities, for using these three cube faces.

Figure 3.2 shows the further restrictions, naturally

arising from these vector cube models. To share the E and B vectors with the Poynting vector edges, we find that only two arrangements (parallel and crossed) are possible. We have anticipated later results and have labeled the parallel Poynting (double edges) as the neutrino type and the crossed double edges as the electron type.

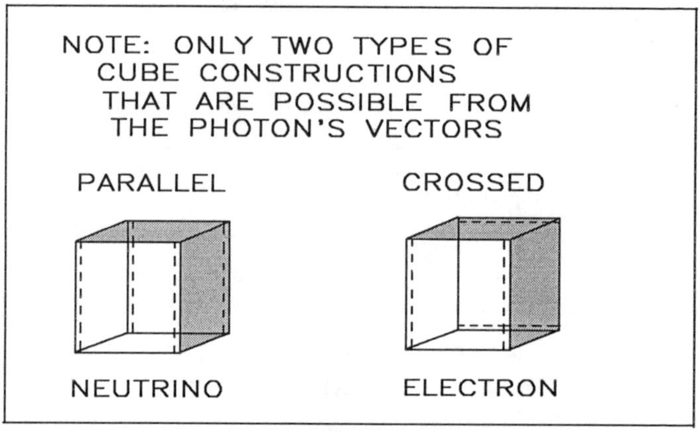

FIGURE 3.2 Natural boundary on edge arrangements.

LEPTONS AT LAST!

When we exhaust *ALL POSSIBLE WAYS* to connect the photons into cube structures we find (*Figure 3.3*) nature is limited to *FIVE* (5) and only five *BASIC* energy constructed particles. We have anticipated later results and have labeled these models as the leptons they mimic. In building the various cube structures, we are restricted to a natural flow, of E and B vector directions, into and out of the Poynting vector (double edges). By inspection, a number of details are apparent. For example, the positron and electron are conjugate structures because there is a E and B vector, on one, where there is a B and E vector, respectively, on the other. This suggests that the original fireball splits apart by breaking up the Poynting vector (photon) edge and forming secondary photons (Poynting vector) cube edges. This would explain conjugate particle pairs always forming, from energy fireballs.

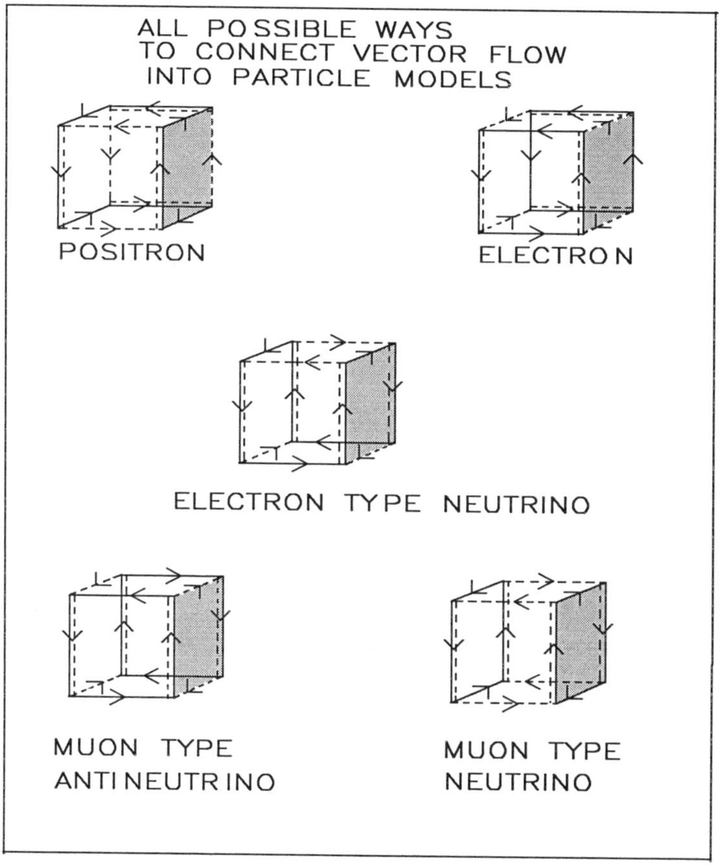

FIGURE 3.3 Natural boundary on basic particle types.

Notice that the electron type neutrino model does not show its conjugate. However, by inspection, it is apparent that the electron type neutrino becomes its own antiparticle when rotated about any axis 180 degrees. This singular result, by a chance arrangement of vectors, is *ABSOLUTELY REQUIRED* later on when we construct the composite proton and neutron models. The muon type neutrino has a unique conjugate structure that matches its mate. The muon type neutrino will prove to create all unstable particles that decay in 2.2 microseconds or less.

MECHANISMS OF CREATION.

So, there we have it. The models show a pleasing closed nature. The coherent modeling process has shown us unique structures with a striking parallel to nature. The picture that emerges, is that two (zero energy) conjugate resonances cooperate to create energy (the photon) and the photon's structure latches up into only three pair of leptons.

We see that nature has natural boundaries in the number of possible ways to construct basic leptons. We showed, by the calculus of related rates, that perfect cubes would form as cube face perimeters stabilize at exactly the velocity of light. At a unique energy, then, it is possible to connect the photon models into perfect cube structures. Energy, anywhere in the universe, naturally and coherently, connects into the same perfect particle structures. The process is mimicked by this model.

In retrospect, this is the way things must happen, viz. *AUTOMATICALLY and PRECISELY.*

Our task, now, is to examine the particle models to see if they can support and justify the known fundamental physical constants. Again, the models do not disappoint us.

THE PRECISE ELECTRON

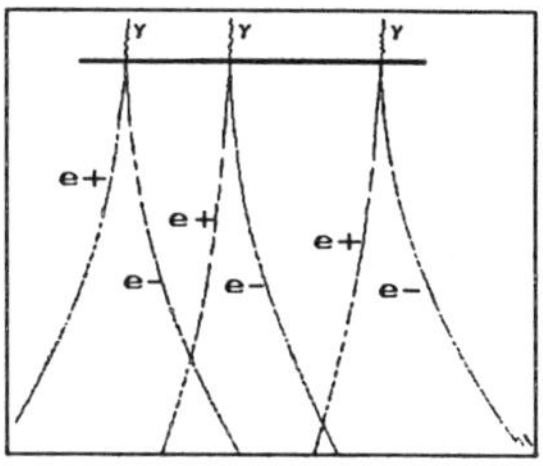

THE PRECISE ELECTRON:

We now proceed by showing that the geometry of the cube can support *ALL* of the electron's fundamental physical constants *EXACTLY*. We know the values of the constants to good precision, and agreement serves to give the model additional credibility. Recent experiments (Dehmelt et. al.) using a Penning trap, have shown that any electron picked at random has exactly the same characteristics (within the experimental accuracy of nine decimal places) as any other.

What mechanism conspires to create the perfection of the electrons? Electrons, as we have seen, are created from photons (Figure 2.0). The electron's precise size cannot result from the photon's energy alone, because the photon can have any frequency whatsoever. The variable frequency (f) gives energy a continuum from ($E = hf$). It must follow that the electron forms, at its precise energy, by some intrinsic quality of the photon's structure. Apparently, the photon is only able to latch up at a one quantum sized chunk of mass energy. We know that all particles heavier than the electron are composites. Demanding a unique energy for basic lepton formation will not prevent other (composite particles) from forming. We will accept the theory of unique energy level for basic lepton formation, without any further speculation, as if it were an aphorism.

In Chapter 3 we developed structures for the subatomic particles from a model for energy. There are only five structures possible, and they constitute three pair of particles, the electron type neutrino being its own anti-particle. Note in *FIGURE 3.3* the candidates for the positron and electron. We now will see that these structures support the known characteristics of mass, charge, spin angular momentum and magnetic moment.

The cube edge length can be deduced from viewing the sinusoidal velocity vector development using the orthogonal vectors as the (x), (y) projections from a rotating phasor (**R**) in the complex plane. *FIGURE 4.0* .

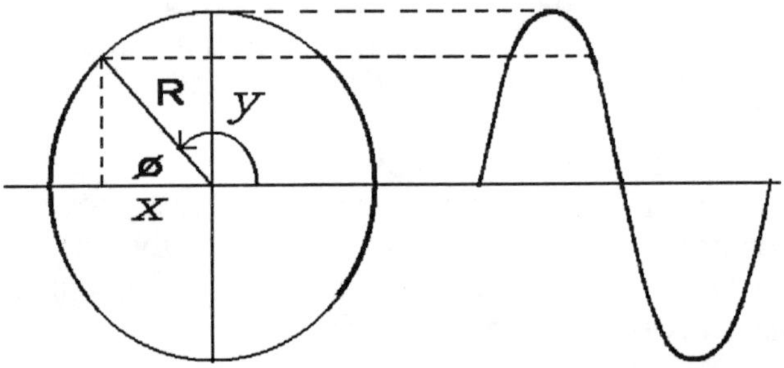

FIGURE 4.0 Sinusoidal projection from complex plane

(y) is the sine and (x) is the cosine function of the projection. The phasor (by convention) rotates counter clockwise. By calculus we can show that if the circumference is set equal to the Compton wavelength (λ) then both (x) and (y) will equal the rationalized Compton wavelength ($\bar{\lambda}$) when (dy/dt) or (dx/dt) equals the velocity of light (c). *FIGURE 4.1* .

Angular velocity $= 2\pi$ f   Freq. f$= c/\lambda$

$$y = \text{R sine } \emptyset \quad \therefore \quad \frac{dy}{dt} = (\cos \emptyset)\,\text{R} \cdot \frac{d\emptyset}{dt}$$

Setting $\frac{d\emptyset}{dt} = 2\pi c/\lambda$, $\frac{dy}{dt} = c$ and

$$x = (\cos \emptyset)\,\text{R} \quad \text{Then } x = \lambda/2\pi = \bar{\lambda}$$

FIGURE 4.1 Calculus of vector lengths from projections

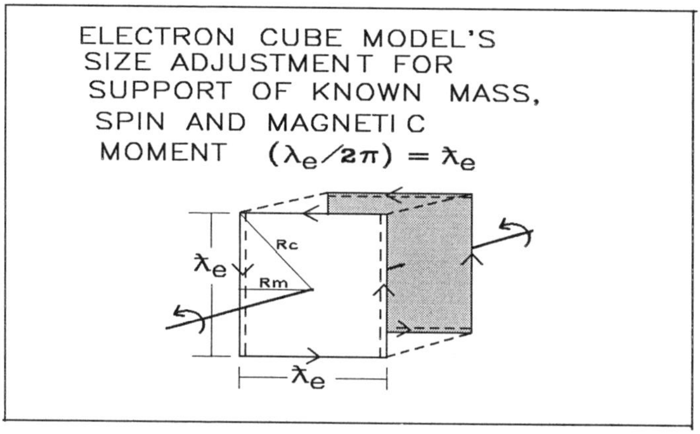

FIGURE 4.2 Electron size geometry.

FIGURE 4.2 is the electron model taken from (*FIGURE 3.3*). Note that the vectors in the front and back cube faces are thrusting their momentum in the same direction (counter clockwise). This is the source of electron spin. This structure is spinning as a natural (chance) arrangement of vectors, with the spin axis going through the center of opposite cube faces. The linear momentum of the photon is structurally conserved in the electron's spin angular momentum. This immediately dictates that the electron model is spinning at (c) to preserve the effective mass (energy) of the photon. Given the mass of the electron, it is a simple, straight forward, exercise to verify that the cube has an edge length equal to the rationalized Compton wavelength as from *FIGURE 4.1* .

For convenience, we now repeat a list of the electron's characteristics here: See also APPENDIX A.

Electron mass (m_e)= 9.1093897 x 10^{-31} kg
Charge (e)= 1.60217733 x 10^{-19} A.s
Vel. of light (c)= 2.99792458 x 10^{8} m.s^{-1}
Spin ($\frac{1}{2}\hbar$)= 5.2728633 x 10^{-35} J.s
Compton wavelength (λ_e)= 2.42631058 x 10^{-12} m
Rationalized ($\bar{\lambda}_e$)= 3.86159323 x 10^{-13} meter

Magnetic moment (μ_e)= 9.2847701 x 10^{-24} J/T
Bohr magneton $\mu_B = \frac{1}{2} (e\, c\, \lambda_e) = A \cdot m^2$
(μ_B)= 9.274015429 x 10^{-24} J/T

The electron's magnetic moment (μ_e) and spin angular momentum ($\frac{1}{2} \hbar$) provide a means to verify that the cube edge length is exactly (λ_e).

SPIN ANGULAR MOMENTUM, ELECTRON:

Spin angular momentum (physically) requires a mass on the end of a radius arm with a certain velocity of gyration. *FIGURE 4.3* .

SPIN = MASS × VELOCITY × RADIUS

$$\frac{1}{2} \hbar = m_e \times c \times Rm$$

$$Rm = \hbar / 2\, m_e c$$

$$\lambda_e = \hbar / m_e c \quad \therefore \quad Rm = \lambda_e / 2$$

FIGURE 4.3 Mass radius of gyration

In *FIGURE 4.2*, the electron's mass radius of gyration (Rm) extends to the mid point of a photon edge. This is as it should be. In a rotating system, it is as if the mass (momentum) was concentrated at a midpoint of an edge, and locates the mass radius. By the laws of right triangles, then, the electron has an edge length of 2(Rm) or exactly (λ_e) as was anticipated (*FIGURE 4.1*).

BOHR MAGNETIC MOMENT, ELECTRON:

We know the magnetic moment of the electron is nearly equal to one Bohr magneton (μ_B), and we can use the Bohr magneton to determine an (idealized) charge radius of gyration (Rc). *FIGURE 4.4* .

$$\mu_B = \frac{1}{2} (e\, c\, \lambda_e) = A \cdot m^2$$

$$A = e\, c / \lambda_e$$

The electron's current A= 19.79633951 Ampere

There are two current loop (m^2). areas in the electron model, the front and back cube faces, as follows:

$$2\,m^2 = \frac{(\lambda_e\ c\ e)\ \lambda_e}{2\ c\ e} = \frac{(\lambda_e)^2}{4\pi}$$

$$m^2 = Rc^2\,\pi = \frac{(\lambda_e)^2}{8\,\pi}$$

$$Rc^2 = \frac{(\lambda_e)^2}{8\,\pi^2} \quad \therefore \quad Rc = \sqrt{\frac{(\lambda_e)^2}{2\cdot 4\pi^2}}$$

$$Rc = \frac{\lambda_e}{\sqrt{2}} \quad \therefore \quad Rc = \sqrt{2}\ Rm$$

FIGURE 4.4 Idealized charge radius.

FIGURE 4.4 calculates the area of the model's two current loops, from the Bohr magneton divided by the electron's current. The area of one loop is obtained as half of the two identical loop areas, and is expressed in terms of the charge radius of gyration (Rc). Note, finally, the charge radius has the required ($\sqrt{2}$) pythagorean ratio to the mass radius (Rm) developed earlier. This clearly shows that the current resides at the model's cube corners, as shown in *FIGURE 4.2* . It has been said that fermions seem to rotate twice, while the world turns once, to account for the magnetic moment 'g' factor of approximately 2. This cube model suggests that the circulating current acts like a two turn coil, and this may explain the paradox.

ANOMALOUS MAGNETIC MOMENT, ELECTRON:

After WWII, the advances in microwave engineering made it possible to measure the magnetic moment of the electron, accurately. It was confirmed that the magnetic moment of the electron was slightly larger than the Bohr magneton. This sparked an ongoing battle between experiment and theory to measure/predict to ever increasing precision. The theory is known as quantum electrodynamics (QED). To claim its agreement with experiment, (QED) must resort to questionable mathematics. For example, (QED) gets rid of infinities that develop at small distances, by the mathematically faulty 'renormalization'. We are not happy with this and it has prevented proving that the theory of quantum electrodynamics is mathematically self consistent.

Rather than use the 'QED self field hypothesis' we will stay with our geometric interpretations and simply enlarge the current loops slightly. We postulate that the current, at the cube corners (*FIGURE 4.2*) protrudes outside of the exacting cube, like an elbow. We will call the protrusion (α_μ). Accepting this, we can calculate a normalized 'size' for the geometric protrusion, based on the measured magnetic moment divided by the Bohr. The new charge radius then becomes the old bare (Rc) times a normalized radius of one plus the distance to the protrusion midpoint ($\alpha_\mu/2$). This is as it should be. In a rotating system, the influence of the protrusion acts from a centroid. *FIGURE 4.5* .

$$\mu_e \neq \mu_B \qquad \mu_e = \mu_B\,(1+\alpha_e)$$

$$Rc = \frac{\lambda_e}{\sqrt{2}}\,(1+\alpha_\mu/2) \qquad \mu_e = 2\,A\cdot m^2$$

$$\mu_e = \left[\frac{2\,c\,e}{\lambda_e}\right]\left[\left[\frac{\lambda_e\sqrt{2}}{4\pi}\right]^2 (1+\alpha_\mu/2)^2\right]\cdot\pi\right]$$

$$\mu_e = \left[\tfrac{1}{2}\,(\lambda_e\,c\,e)\right](1+\alpha_\mu/2)^2 \quad \text{MODEL}$$

$$\mu_e = \left[\tfrac{1}{2}\,(\lambda_e\,c\,e)\right](1+\alpha_e) \quad \text{NIST}$$

$$(1+\alpha_e) = (1+\alpha_\mu/2)^2$$

$$\alpha_\mu = 2\left[\left[\sqrt{1+\alpha_e}\,\right]-1\right]$$

FIGURE 4.5 Anomalous magnetic moment protrusion

Finally, notice the MODEL does derive the Bohr magneton, and this is multiplied by the normalized charge radius plus its protrusion mid point, the quantity squared. Compare this with the NIST expression for the electron's magnetic moment from the Bohr. Finally, we can equate the two 'g' factors and thus calculate the protrusion (α_μ) dimensions (normalized).

POSITRON MODEL SIZE:

The positron, based on Penning trap experiments, is identical to the electron, differing only in having an opposite charge. The positron model (*FIGURE 3.3*) shows a conjugate structure to the electron model. Note the crossed photon vectors, in the front and back cube faces that makes the positron spin, also, causing spin angular momentum, mass and magnetic moment.

WHAT IS MASS?

Rest mass must result from the photon's momentum (effective mass times velocity). We calculated the electron's size in *FIGURE 4.3* and showed that the electron can be modeled as a cube with an edge length equal to the electron's rationalized Compton wavelength. We showed that the spin angular momentum is stored energy in (Joule seconds) after the fashion of energy stored in a classical flywheel. We now ask the question, what part of the electron's rest mass energy is stored in the spin? *FIGURE 4.6* shows the calculations for electron (model) mass (energy).

$$\tfrac{1}{2}\hbar = \frac{J\cdot s}{4\pi} = Rm \times m_e \times c$$

$$s = \lambda_e / c$$

$$Rm = \lambda_e / 4\pi$$

$$\frac{J\cdot\lambda_e}{4\pi\cdot c} = \frac{m_e\cdot c\cdot\lambda_e}{4\pi}$$

$$J = m_e\cdot c^2$$

FIGURE 4.6 Proof that mass (energy) reguires spin.

The spin angular momentum has the dimensions of Joule seconds divided by 4pi. There are two unknowns, the energy (J) in Joules and the time (s) in seconds. The electron's time period (s) is equal to its Compton wavelength divided by the velocity of light (c). The electron's mass radius (Rm) we know from the vector model. Inserting the relationship for time (s) and (Rm) and transposing, finally obtains the Einsteinian for the electron. We see that *ALL* of the electron's rest mass energy is stored in the spin angular momentum. We conclude that rest mass requires spin to preserve the photon's effective mass and momentum, and further, then, that zero spin particles can only be clumps of spin 1/2 particles. (i.e. like the helium four nuclei that exhibits zero spin statistics but is, in reality, a bound state between two protons and two neutrons, whose spins buck out.)

SUMMARY:

The model continues to show a coherence by supporting the known characteristics of the basic electrons.

In the next chapter we will discuss the neutrino model, current neutrino theories and experiments to detect the 'ghostly' neutrinos.

THE NEUTRINO AS A NEW CLASS OF PARTICLE

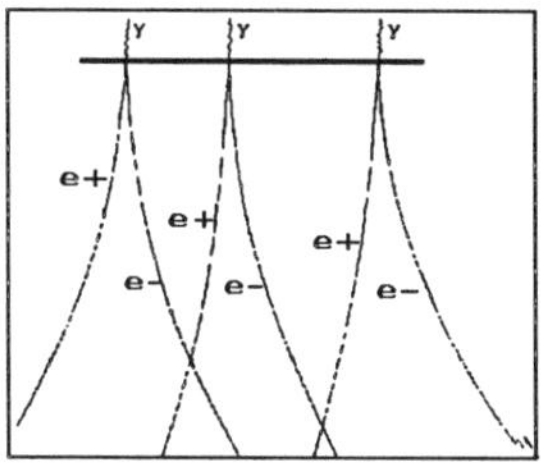

The neutrino, as developed in the vector models, will seem bizarre to those readers versed in current neutrino theory. The vector neutrinos (CHAPTER 6) will be shown as providing access to the nested cubes of multidimensional space. To access subspace, the vector neutrino, on absorbing energy, must become smaller than the ordinary three dimensional space set by the quantum sized electron. The vector neutrino of this model cannot translate when absorbing energy, but must vibrate along all dimensions alike. Arguments, for these processes, proceed by first reviewing Dirac's theory about the mechanism for the motion of electrons.

ZENO'S PARADOX OF THE ARROW:

Man has been puzzling over the paradox of motion for at least the 2500 years since Zeno of Alea posed the paradox of the arrow. It goes something like this:

"An arrow at rest occupies a space equal to its volume. Likewise, a flying arrow, at any instant, occupies a space equal to its volume. Motion is impossible to describe!"

Dirac, in his 1932 Nobel prize lecture, touched on a possible answer to the Zeno paradox. Quote:

It is found that an electron which seems to us to be moving slowly, must actually have a high frequency oscillatory motion of small amplitude superimposed on the regular motion which appears to us. As a result of this oscillatory motion, the velocity of the electron at any time equals the velocity of light. This is a prediction which cannot be directly verified by experiment, since the frequency is so high and its amplitude so small.

Dirac's theory, for the mechanics of motion, can now be modeled using the vector structures developed for energy and matter (CHAPTER 2 and 3). It was proposed, in CHAPTER 2, that the photon model comes in two styles. In one case the Poynting vector was pushed ahead, in the other, the Poynting vector was trailing the E and B vectors. In both cases the energy consists of two conjugate (zero energy) resonances which, taken together, create motion (as translation) from one stationary point, in their defined space, to another. The photon, alone, has the mechanism that produces motion. Matter does not move unless energy acts on it. *THE PHOTON'S ABILITY TO MOVE MUST (SOMEHOW) BE TRANSFERRED TO MATTER.* What change must take place to cause matter to move? We have known since Heisenberg (1901-1976) that the photon mediates forces between particles. What has been lacking is a mechanism for *both ATTRACTION and REPULSION*, mediated by the photon. We now have a model for the structure of energy (the photon) and a model for the structure of matter (particles) with which to deduce a theory for motion caused by the photon.

ATTRACTION AND REPULSION:

It is easy to visualize repulsion as the exchange of momentum mediated by a photon exchange. Two billiard balls hitting together, compressing their near sides, and expanding apart, is a classical example. But, what is the mechanism of attraction, for example, between a positive positron and a negative electron? A theoretical means is shown in (*FIGURE 5A*). We postulate that attraction is caused by the exchange of conjugate photons, which act on the receiving particles by compressing their structures from the far sides, so that they each expand toward the other. On the other

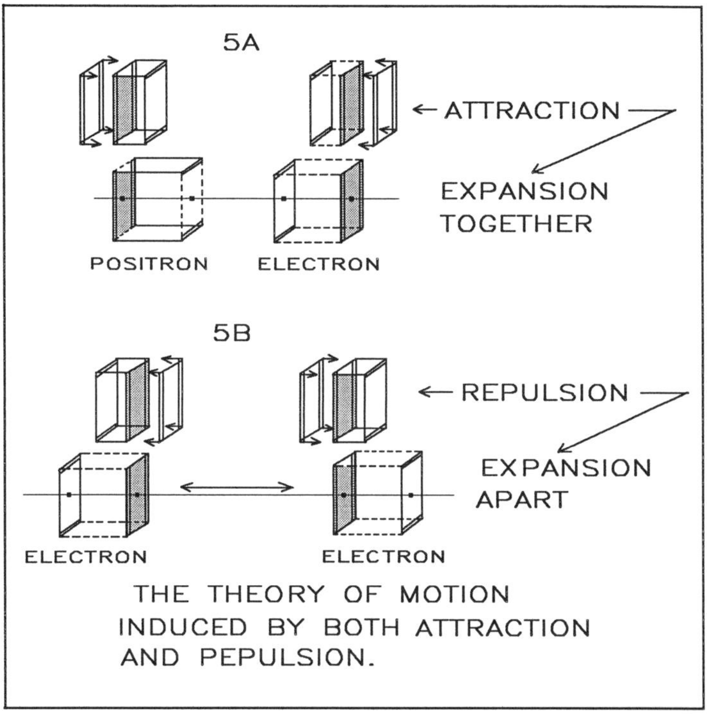

FIGURE 5 Dirac's motion theory vector model.

hand, in (FIGURE 5B) repulsion between negative electrons (or positive positrons) is caused by exchanging like photons that compress their structures from the near sides, so that they expand apart. Note the vector structure of the positron and electron have all axial vectors (*FIGURE 5.1*). The vectors, along the axis of spin, are of either magnetic B or electric E type. It is imagined that being the same type allows the axial expansion and contraction necessary for translation (motion). Electron beams at close to the velocity of light (as in accelerators) are all polarized by spins. This tends to prove the Dirac theory that electrons move by axial expansion and contraction, inch worm style under the influence of photons. Inertia, then, is the resonant "sloshing back and forth" of the energy in matter structures. 200 years after Newton's laws for inertia, we just now have a possible model for inertia, based on the vector models.

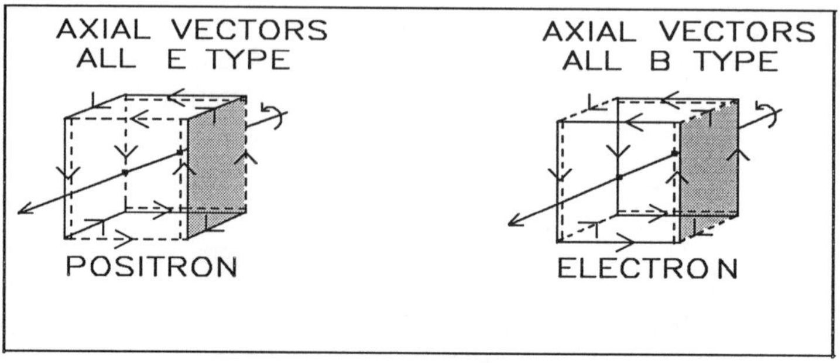

FIGURE 5.1 Positron and electron vector structures.

THE INCREDIBLE NEUTRINOS:

Lets step back and look at the history of the neutrino. Pauli (1900-1958) postulated a neutral particle (in 1930) to explain the apparent failure of energy conservation in the beta decay of radioactive nuclei. With the detection of the neutron (in 1932) by Chadwick (1891-1974) beta decay became better understood and Fermi (1901-1954) used the neutron decay into an electron (beta particle) and neutrino as the basis for a theory of weak interactions (in 1933). Unfortunately, Fermi drew a parallel between the neutrino and the photon. Bear in mind that the neutrino had not been detected, and Fermi's neutrino paper was based on the (well known) electromagnetic emissions, and the quantum theory of radiation using the creation and destruction operators of Dirac-Jordan-Klein-Wigner. The paper was so speculative that the Editor of NATURE refused to print it, and the paper was first printed in Italian in *Zeitschrift fur Physic* (in 1933). Fermi never published anything else on the subject, yet his theory forms the basis for all present day "Weak Decay theories".

For 25 years no claims for neutrino detection could be

made. Everyone thought that neutrinos should have zero rest mass (but, today, the jury is still out on neutrino mass). Everyone thought that neutrinos travel at the velocity of light and could penetrate light years of lead (the velocity of light was based on the idea that if we assign zero rest mass then we have to assign the velocity of light on the basis of relativity) of course if the neutrinos have mass............all bets are off. So, for 25 years the neutrino was sought in the cloud chamber records among the cosmic ray artifacts. Not a single neutrino was found, and the lack of neutrino detection *ASSUMED* neutrinos were "weakly" interacting particles. Lower limits for neutrino detection were estimated from the negative (lack off detection) results and resulted in the famous "light years of lead" penetrating ability. This, of course, is circular reasoning based on the assumption that neutrinos are exempt from the wave particle duality of both energy and matter, and somehow penetrate large distances in matter.

After WWII the first "Big Science" experiment was using the Savanna River Reactor neutron flux to "theoretically" produce a large flux of neutrinos for possible detection. (Remember, the assumption was that neutrinos ought to travel at the speed of light and penetrate light years of lead.) The new theory for neutrino detection was based on the idea of "inverse beta decay". Inverse beta decay had been detected by nuclei absorbing an electron, but never by the nuclei absorbing a neutrino. Never the less, the experiment was setup on that highly speculative idea of inverse beta decay. Today, the literature is so cluttered with claims for detecting the neutrino that detecting the neutrino by inverse beta decay is taken for granted. *ACTUALLY, THE NEUTRINO HAS NEVER BEEN DIRECTLY OR UNAMBIGUOUSLY DETECTED*. Let's look at some of the more well known experiments, and you will see how tentative the claims are.

(EXAMPLE 1) The first tentative claims for detecting

the neutrino were based on detecting inverse beta decay at the fission reactor at Hanford (1953). Now, this approach was highly speculative because inverse beta decay, induced by the neutrino, had never been detected in the thousands of comic ray cloud chamber photographs. Inverse beta decay had been seen induced by the electron, so it was assumed that the neutrinos *OUGHT* to also. The experiment was run with the expectation that a proton could absorb an antineutrino, forming a neutron and emitting a positron. The positron was then to annihilate with an electron and produce two photons of the characteristic 0.51+ MeV energy. Because of the large background from cosmic rays and strays from the reactor, the test was inconclusive. The signal was just about the same with the reactor off. Later, (1956 to 1958) improved experiments were performed at the Savannah River Reactor. Again it was impossible to get rid of background events, and required running the detectors with the reactor off for 24 hours and comparing the count to that with the reactor on. Out of the hundreds of background events, the experimenters claimed 70 fewer events with the reactor off.

(EXAMPLE 2) The Homestake Mine in Lead, S.D. experiment has been operating for about the last 25 years. The detector is a tank filled with 100,000 gallons of tetrachloroethylene (C_2Cl_4), an ordinary cleaning fluid. Again the detection depends on the (never proved) inverse beta decay by neutrinos. The inverse beta decay is supposed to transmute the Chlorine-37 into radioactive Argon-37. This reaction also can be mimicked by high energy muons, from cosmic rays, knocking protons from nuclei, if the proton enters the detector. Needless to say, the experiment is not detecting the number of "events" predicted by the theoretical solar neutrinos.

(EXAMPLE 3) The experiment at Brookhaven National Laboratory, Upton, N.Y. to test for two kinds of neutrino in (1962) was also based on the mistaken belief that muons could be created by a muon neutrino going into a neutron or proton. Again, these interactions were speculative, having never before been detected. The experiment used a spark chamber gated by a Cerenkov counter exposed to the pions produced by 15 BeV protons striking a beryllium target. The gating scheme was an attempt to reduce the background muons from cosmic rays. *THE CORRECT BEAM PHASING WAS VERIFIED BY RAISING THE MACHINE ENERGY TO 25 BeV* and counting the muons that then penetrated the 13.5 meter iron shield. (This muon signal is coming from the same direction that is later claimed to be caused by neutrino induced muons!!!!!) Also, muons, from cosmic rays, which stop in the spark chamber, can and do, simulate neutrino events. In their efforts to discount the cosmic rays muons, they offered the following arguments. We quote verbatim from the Physical Review Letters, Volume 9, Number 1, July 1, 1962. Quote:

1. The "events" are not produced by cosmic rays. *Muons from cosmic rays can and do simulate neutrino events. This background is measured experimentally by running the AGS machine off on the same triggering arrangement except for the Cerenkov gating arrangement. The actual triggering rate then rises from 10 per hour to 80 per second. In 1800 cosmic ray photographs thus obtained, 21 could be accepted as neutrino events. Thus 1 in 90 cosmic ray events is neutrino like. Cerenkov gating and the short AGS pulse effect a reduction by a factor of (about) 10^{6} since the circuits are "on" for only 3.5 microseconds per pulse. In fact in the body of the data represented by Table I, a total of 1.6 x 10^{6} pulses were counted. The equipment was therefore sensitive for a total time of 5.5 sec. This should lead to 5.5 x 800 = 440 cosmic ray tracks which is consistent with observation. Among these there should be 5 ± 1 cosmic ray induced "events". The remaining 51 events cannot be the result of cosmic rays.* End Quote.

Those "51 events" are simply muons coming from the direction of the AGS accelerator. *THERE IS NO WAY TO UNEQUIVOCALLY BLAME THOSE MUONS ON*

SOME THEORETICAL REACTION BY A NEUTRINO BECAUSE MUONS ARE PRODUCED NORMALLY BY THE 15 BeV PROTONS STRIKING THE BERYLLIUM TARGET. IN FACT, REMEMBER THAT THE GATING WAS CALIBRATED BY RAISING THE AGS MACHINE ENERGY TO 25 BeV.

(EXAMPLE 4) Probably the most unbelievable were the claims for detecting neutrinos from the 1987A supernova at Kamiokande II and IMB detectors set up for proton decay experiments. The claims are based on the theory that the neutrino can somehow "knock on" an electron. Also the theory of inverse beta decay was thought to be able to produce electrons. Accelerated electrons can be detected by Cerenkov light cones in the large tank of water. But, electrons produced by neutrinos have never been seen in comic ray photographs, so the premise is highly speculative.

It is estimated that about 10^{58} neutrinos with approximately 10 MeV energy would be produced in the supernova 1987A. Kamiokande II claimed to have seen, over a two second period, 9 electrons of energies, 20,14,8,9,13,6,35,21, and 20 MeV. Note that only three of these are of the right energy (i.e. less than 10 MeV). And, only two "events" pointed in the right direction!!!! This is not very conclusive evidence that neutrinos came out of all that supernova matter, traveled the 50 kiloparsec to a minor planet and decided to react in a small tank of water in a deep mine. The more plausible scenario is that cosmic ray muons (they detect an average of three muons a second in Kamiokande II) are producing the electrons when they decay with an average energy of 50 MeV. Such muon decay electrons could loose energy and produce the "events" rather than 10 MeV neutrinos.

THE VECTOR NEUTRINOS:

The vector neutrinos are constructed with all four

Poynting vector (photon) edges parallel. If we follow the development of direction in the Poynting vectors (*FIGURE 5.2*) the total momentum bucks out. The neutrino model seems dead and motionless. The structure is not spinning like the electron or positron. (The neutrinos will later prove to spin only when in concert with the electron or positron in composite particles).

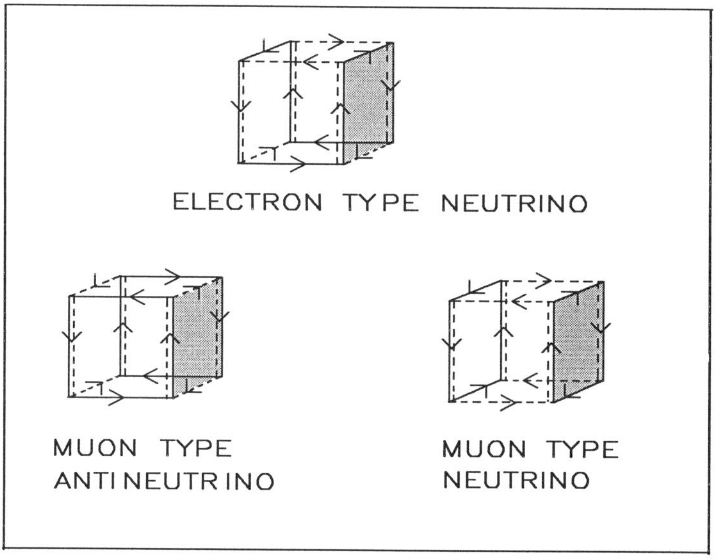

FIGURE 5.2 Electron and muon type vector neutrinos

The neutrinos are not charged and hence do not exchange the electromagnetic photons, as does the electron and positron (*FIGURE 5*). When a photon attaches to the neutrino as to move it, the bucking structure is postulated to store the energy by shrinking the structure rather than move the structure. (*FIGURE 5.3*) The Compton wavelength (λ_c) has mass (M) in the denominator, so more mass (energy) means a smaller sized neutrino cube. This gives the possibility to form the neutrino inside of the electron's Compton wavelength (λ_e) vortex, in order to form particles heavier than the quantum. *THE NEUTRINO IN THIS MODEL BECOMES A NEW CLASS OF PARTICLE, THAT, RATHER THAN*

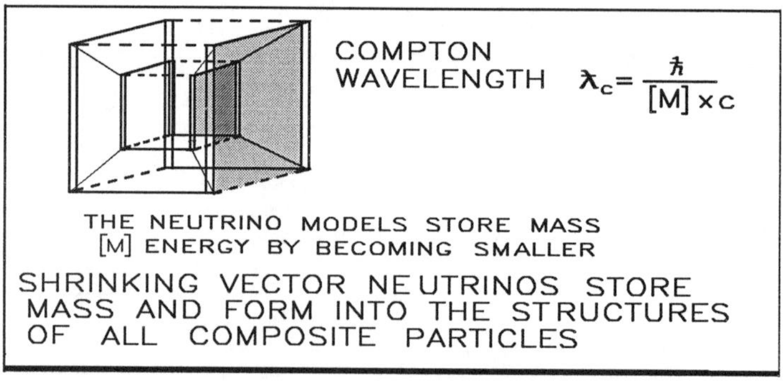

FIGURE 5.3 Vector neutrino model absorbing energy.

MOVE WHEN ABSORBING ENERGY, MERELY GETS SMALLER AND GIVES ACCESS TO THE NESTED CUBES OF MULTIDIMENSIONAL SPACE. Were it not so, the universe would be composed of nothing but minimum mass (energy) leptons.

In the next chapter we will use the special characteristics, deduced for the vector neutrinos, to form the composite proton, neutron and muon.

NOTES: Recent experiments in Canada, at the University of Guelph, (Simpson) seem to indicate that the neutrino may have a mass of a few KeV. Experiments at Lawerence Berkeley Laboratory (Norman) tend to also verify a similar neutrino mass.

If a free neutrino is spinning and has mass, then it should, of course, also exhibit a charge and a magnetic moment and be detected in ionization detectors. The neutrino is not detected, so the vector model's indications that the neutrino can only spin when in concert with the electron, and cannot move nor spin when free, is the only possible explanation for a "ghostly neutrino" that stores mass (energy).

Further, the experimental mass (energy) would effectively prevent the neutrino from traveling at the velocity of light and penetrating "light years of lead" .

COMBINING LEPTONS

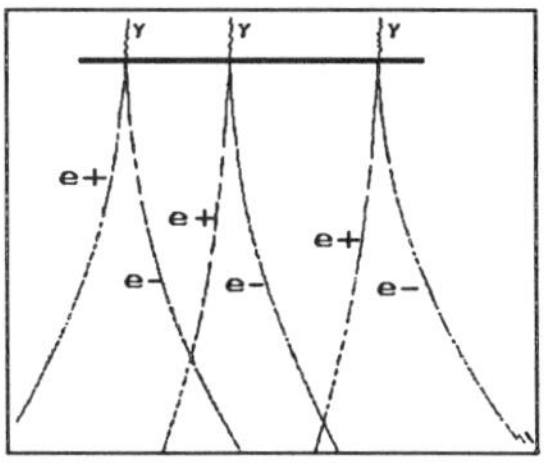

Energy is limited to structuring leptons (electron pair, electron type neutrino pair and the muon type neutrino pair) as demonstrated in (CHAPTER 3). This being the case, we are forced to construct the proton and neutron from some combinations of leptons. To build these composites (i.e. the proton and neutron) a starting point would be to review what experimental results indicate about their composition. All composite particles decay (except the stable proton) ultimately, into leptons. Any excess energy, from the decay, goes into photons or kinetic energy of the decay products. The proton, even though stable against decay, is believed to be a composite because its magnetic moment does not involve all of the proton's structural mass. Further, the proton, in what is known as 'K' shell electron capture, can change into a neutron. Conversely, the neutron beta decays by returning to a proton, an electron, and a neutrino.

The picture that emerges is that leptons go into nucleons (protons) and come out of nucleons (neutrons). The obvious conclusion is that composite particles are composed of their lepton decay products.

LEPTON COMBINATION RULES:

In CHAPTER 5 we developed the premise that the neutrino was a new class of particle that absorbed and stored (mass) energy by size reduction *(FIGURE 5.3)*. The quantum mechanics can only make basic particles as Planck sized chunks of (mass) energy, initially, so the neutrino's size reduction allows the neutrino to form inside of the electron's or positron's spinning vortex. The breakthrough with this modeling process came by deducing the effect of the neutrino model inside of the electron or positron. It is possible to arrange the neutrino inside of the electron (or positron) and simply add vectors, member by member. The vector combination rules are shown in *FIGURE 6.0* for one face

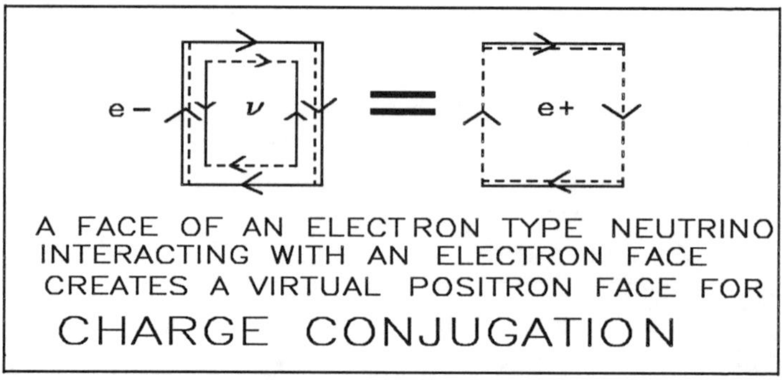

FIGURE 6.0 *Vector combination rules.*

of the cubes, and are straight forward. Note that *LIKE* vectors going in the opposite direction cancel, and *UNLIKE* vectors going in the same direction form a virtual photon (Poynting vector) cube edge. The face of an electron type neutrino combines with an electron face to form a virtual positron face. Note in *FIGURE 6.0* that the vectors combine to spin up the assemblage of electron and neutrino. The model indicates that the neutrino can only spin when in concert with a host electron.

CHARGE CONJUGATION:

The arrangement in *FIGURE 6.1* of an electron type neutrino inside of an electron creates a virtual positron. The reader can review the vector structures of *FIGURE 3.3* and verify that the electron type neutrino does, indeed, combine (vector by vector) with an electron to form a virtual positron (charge conjugation!). The virtual positron then spins up the assemblage.

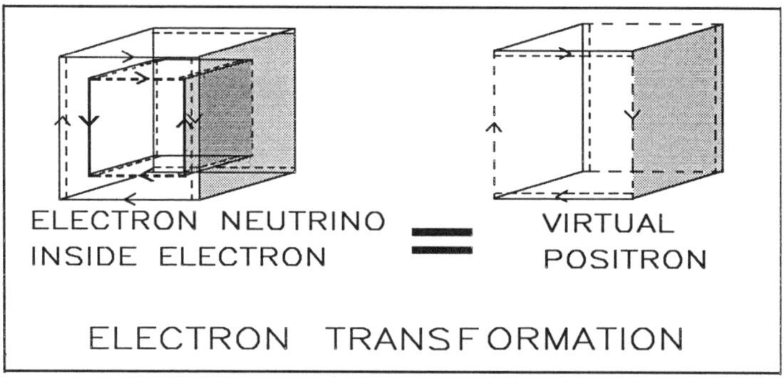

FIGURE 6.1 Electron to virtual positron (conjugation)

The electron type neutrino is its own antiparticle and *FIGURE 6.2* shows the orientation of neutrino inside of the positron (real or virtual).

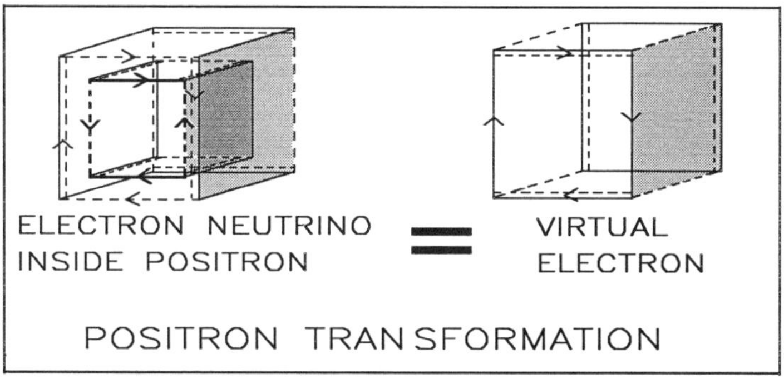

FIGURE 6.2 Positron to virtual electron (conjugation)

The vectors add up to a virtual electron. This ability of the electron type neutrino to cause charge conjugation does not violate charge conservation because the neutrinos are created in pairs inside of the vortex. Acting in pairs, the neutrinos always return to the same polarity as the original type of virtual particle, that picked up the pair, thus assuring charge conservation. The ability of the electron type neutrino to cause charge conjugation stores electrical potential energy in the space between the activated neutrino and host assemblage. This energy is calculated by the well known equation: *(FIGURE 6.3)*.

$$\text{ENERGY} \quad J = \left[\frac{e^2}{4\pi \varepsilon_0 d}\right]$$

FIGURE 6.3 Electrical potential energy from charge conjugation between nested cubes.

Where (d) is the separation distance in meters and the other symbols have the usual MKSA meaning (TABLE 1, APPENDIX A).

SCALING INNER VECTOR NEUTRINOS:

The cube geometry allows us to *PERFECTLY* and *AUTOMATICALLY* scale to the mass of the composites. The vector electrons are spinning by virtue of the crossed Poynting vector edges in their structures *(FIGURE 5.1)*. The vector neutrinos are not spinning by virtue of their parallel Poynting vector edges *(FIGURE 5.2)*. The vector neutrinos, however, unlike the electrons, have the ability to store energy (locally) by alternately becoming smaller along all edges alike *(FIGURE 5.3)*. We immediately can visualize how nature forms composite particles anywhere in the universe with stunning precision *(FIGURE 6.4)*.

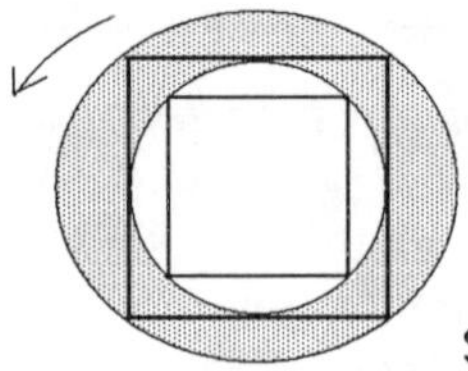

FIGURE 6.4 Rotating positron creates a vortex that only accepts a precisely scaled non spinning neutrino with the proper energy content.

The neutrino forms in the vortex of the spinning positron

(or electron) and is spun up when the vector neutrino can just make an interference fit in the vortex as shown. The antineutrino then absorbs enough energy to just fit in the resulting assemblage's vortex. Scaling is automatic and precise as shown in *(FIGURE 6.5)*.

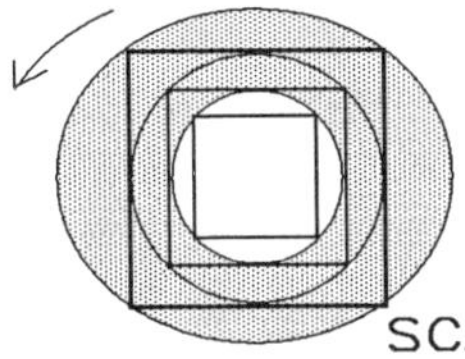

FIGURE 6.5 Virtual electron assemblage creates a vortex that only accepts a perfectly scaled non spinning neutrino with proper energy content.

The cubes edge lengths are reduced, in the vortex scaling, by the small (precise) elbows that enlarge the current loops (slightly) to account for the anomalous magnetic moment *(FIGURE 4.5)*. For precise scaling, then, we must account for this (α_μ) protrusion at the corners. Two protrusions act along each edge, reducing the Compton wavelength and therefore causing a small increase in the mass stored in each inner vector neutrino. A convenient (normalized) adjustment becomes: *(FIGURE 6.6)*.

(a) $2\alpha_\mu(\sqrt{2}/2) = \sqrt{2}\,\alpha_\mu$

(b) $\left[(\sqrt{2}/2)^N - \left[(\sqrt{2}/2)^N(\sqrt{2}\,\alpha_\mu)\right]\right]$

(N = INNER SITE NUMBER)

FACTORING THIS EQUALS:

(c) $(\sqrt{2}/2)^N\left[\,1 - (\sqrt{2}\,\alpha_\mu)\right]$

FIGURE 6.6 Scaling factor adjustment for the reduced edge length due to the anomalous magnetic moment protrusions.

In *FIGURE 6.6* (a) we calculate the two protrusions that act on each inner edge to reduce their length, slightly. In (b) we develop a general expression for a normalized edge length reduced by the two protrusions of (a). The square root of two divided by two to the "N" power is the normalized edge length of any "Nth" inner site. Factoring out the scaling factor, finally gives the general expression for the scaling factor of any inner site "N". Combining a normalized scaling factor with the normalized protrusion adjustment, as deduced in *FIGURE 6.6*, we can scale inner vector neutrino edge lengths (L_n) from the electron's rationalized Compton wavelength ($\lambda\!\!\!^{-}_e$) as: *(FIGURE 6.7)*.

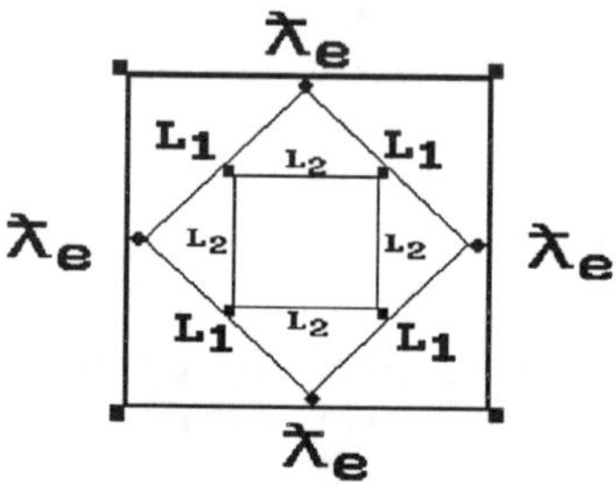

$$L_N = \lambda\!\!\!^{-}_e \left[\left[\sqrt{2}/2\right]^N \left[1-\sqrt{2}\,\alpha_\mu\right]\right]$$

FIGURE 6.7 Edge length of any internal cube obtained by powers of the normalized scaling factor, reduced by two protrusions, times the electron's rationalized Compton wavelength.

When each inner vector neutrino's edges are parallel (and interacting) charge conjugation, (at each inner structure), will store energy (as electrical potential) *(FIGURE 6.3)*. This energy adds to the rest mass of the composite particle and must also be taken into account. To calculate the electrical potential energy given in *(6.3)*, we must determine a normalized separation distance (d) between inner cubes, as follows: *(FIGURE 6.8)*.

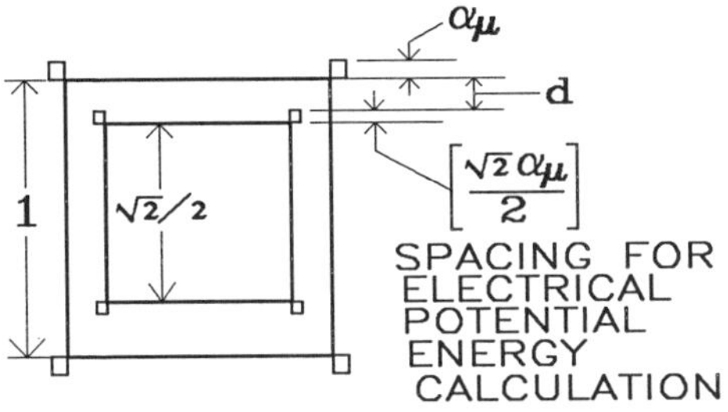

$$d = \left[\left[\frac{(1 - \sqrt{2}/2)}{2} \right] - \left[\frac{\sqrt{2}\,\alpha_\mu}{2} \right] \right]$$

FIGURE 6.8 Normalized separation calculation.

SUMMARY:

The model has been allowed to dictate the direction we have taken, without regard to any conflicting dogma. This is as it should be. The results can only be a natural, orderly, progression of revelations (not forced as in the standard model). We now have reason to believe that only three pair of particles can be directly constructed by energy on account of the quantum. It would appear that matter would be forever confined to the dimensions of ordinary space and time, set by the quantum, except for the vector neutrino's ability to entrain energy and shrink to sub space size. The composite particles, then, are composed of the nested cubes of multidimensional space. It is the neutrino that (happily) is able to give us access to other dimensions of space and time, and thus provides a means to continue the models coherently. The picture that emerges is that nature is able to form matter, *AUTOMATICALLY and PRECISELY*, anywhere in the universe. The laws of chance, which exhaust all possible ways to connect energy to itself, assures that the same particles form universally. In the next chapter we will use the deduced relationships for scaling to construct *PERFECT* models for the proton and neutron.

THE PROTON AND NEUTRON MODELS

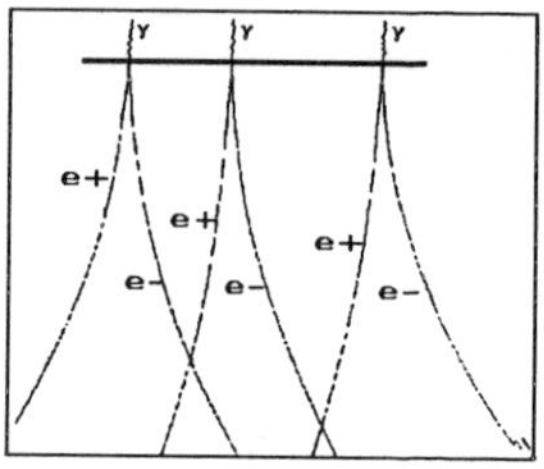

PROTON CONSTRUCTION:

In a composite particle model, we found that each inner vector neutrino can be perfectly scaled, as was shown in *(FIGURE 6.7)*. As each inner vector neutrino (cube) entrains energy and scales down, to an interference fit in the vortex of the outer spinning assemblage, the vectors add, creating virtual photon edges. These virtual photons then spin up the total structure. We see that the process proceeds entirely automatically and precisely (as it must).

The proton model forms in the *VORTEX* of a spinning positron, the antiproton in the vortex of a spinning electron. The electron type neutrino, that forms in their vortex, is its own antiparticle and serves to form both proton and antiproton. The models require nine pair of electron type neutrinos to obtain the proton's known mass. The electrical potential energy, that is created between conjugating layers, adds a total of about 5 percent to the proton's total mass, obtaining the CODATA value to within about 3 parts per million. The stability of the proton (or antiproton) is the result of the electrostatic forces that hold the conjugating layers together. The proton model's core particle is a cube with an edge length of 0.7529 x 10^{-15} meters. This compares favorably with Hofstadter at SLAC (*Phy. Rev.* 102, 851 1956) who obtained a charge radius of 0.75 $\pm$ 0.24 (10^{-15}m). Further, the model's core particle, obtained by the scaling, is the correct order of magnitude to predict the known magnetic moment of the proton.

The original model calculations used a spread sheet with every part of the proton model building on the previous calculation. The spread sheet took six hours to calculate with a hand held calculator (TEXAS INSTRUMENTS SR50). The calculations are greatly simplified, by reducing the spread sheet to a summation and using a computer. We start with an expression that represents the mass ratio of the outer positron, which is one, plus the electrical potential energy mass ratio to the next inner site. *(FIGURE 7.0)*

$$e=1+\left[\frac{e^2}{\left[m_e c^2\, 4\pi\, \varepsilon_0\right] \lambda_e \left[\left[\frac{1-\sqrt{2}/2}{2}\right]-\left[\frac{\sqrt{2}\;\alpha_\mu}{2}\right]\right]}\right]$$

FIGURE 7.0 Normalized mass ratio for scaling to the mass ratio of the proton and neutron.

In *(FIGURE 7.0)* the added mass energy is due to the electrical potential energy between the outer positron and the first inner neutrino. In *(7.0)* we have divided the potential energy by the Einsteinian ($m_e \cdot c^2$) energy to obtain the dimensionless mass ratio. The mass of each inner site, then, is taken as an exactly scaled (and accommodated) version of the outer positron (or electron for the antiproton).The scaling factor is the inverse of *(FIGURE 6.6(c))* to give the mass ratio of the proton. *(FIGURE 7.1)* gives the scaling factor (S).

$$S=\left[\left[\sqrt{2}/2\right]^N \left[1-\sqrt{2}\,\alpha_\mu\right]\right]^{-1}$$

FIGURE 7.1 Universal scaling factor for mass ratio calculations in proton and neutron models.

Putting the relationships developed in Chapter 6 and this chapter, into a summation, finally produces the mass ratio of the proton as in *(FIGURE 7.2)*.

$$m_p/m_e = e + \sum_{N=1}^{18} s \cdot e$$

FIGURE 7.2 Proton model summation truncated after adding nine pairs of scaled vector neutrinos. This exponentially increasing function stops after adding the last 28 percent as a core particle!

The reader can verify that the algebra is correct. The summation in *(FIGURE 7.2)*, using the terms from *(FIGURE 7.0)* and *(7.1)*, can be set up with any scientific mathematics software. The BASIC version in the APPENDIX B will run on most computers also. A sample calculation of the proton's structure is given in TABLE 2. The site number (N) is assigned to the inner neutrinos starting with (1) for the first inner neutrino. Note that the electrical potential energy (mass) contributes 87.61862892 electron masses to the proton. This is a full 5 percent, and the model predicts a mass ratio, for the proton, as 1836.1467745836. This prediction is within about 3.24 ppm (parts per million) of the CODATA recommendation of 1836.152701. (To put a part per million into perspective, a part per million is like measuring the length of a football field to within the thickness of one sheet of paper).

Notice, in TABLE 2, that the first two layers of the proton model add up to exactly the difference in mass ratio between the proton and neutron. (i.e. 2.537628381 electron masses). If we postulate unformed pairs, at outer and first inner site, we immediately have the known beta decay products of the neutron (n) when it becomes a proton (p^+). (i.e. an electron, e^- and electron type neutrino, ν) *(FIGURE 7.3)*

$$n \rightarrow p^+ + e^- + \nu$$

FIGURE 7.3 Neutron decay products predicted by the vector models, and known experimentally

TABLE 2 CALCULATION OF THE PROTON'S AND NEUTRON'S STRUCTURE

SITE N	STRUCTURE MASS RATIO		ELECTRIC POT MASS RATIO		TOTAL MASS RATIO	
OUTER	1.000000000	+	0.050109933	=	1.050109933	
1	1.416535927	+	0.070982521	=	1.487518448	
2	2.003284319	+	0.100384444	=	2.103668763	
3	2.833071853	+	0.141965042	=	2.975036895	
4	4.006568638	+	0.200768888	=	4.207337526	
5	5.666143707	+	0.283930084	=	5.950073791	
6	8.013137277	+	0.401537775	=	8.414675052	
7	11.332287414	+	0.567860168	=	11.900147581	
8	16.026274553	+	0.803075550	=	16.829350104	
9	22.664574828	+	1.135720335	=	23.800295163	
10	32.052549107	+	1.606151101	=	33.658700208	
11	45.329149656	+	2.271440670	=	47.600590326	
12	64.105098214	+	3.212302202	=	67.317400416	
13	90.658299311	+	4.542881340	=	95.201180651	
14	128.210196428	+	6.424604403	=	134.634800831	
15	181.316598623	+	9.085762680	=	190.402361303	
16	256.420392855	+	12.849208807	=	269.269601662	
17	362.633197245	+	18.171525361	=	380.804722606	
18	512.840785711	+	25.698417614	=	538.539203325	
	--------------		--------------		------------	
TOTAL	1748.52814567	+	87.61862892	=	1836.14677458	PROTON
				=	1838.68440296	NEUTRON

The mass ratio, for the neutron, then, is given by doubling up on the two outer sites. The summation equation for the neutron becomes *(FIGURE 7.4)*.

$$m_n/m_e = 2e + 2e\left[\left[\frac{\sqrt{2}}{2}\right]\left[1-\sqrt{2}\alpha_\mu\right]\right]^{-1} + \sum_{N=2}^{18} s \cdot e$$

FIGURE 7.4 Neutron model's summation equation that adds the decay products to the proton model's summation equation.

(FIGURE 7.4) gives the neutron model's mass ratio as 1838.6844029647 electron masses. The model's prediction is within about 400 ppb (parts per billion) (or 0.4 ppm) of the CODATA recommendation. Coincidences continue with this vector model. We now have the only successful model for, not only giving the mass of the proton, but the mass *AND EXACT DECAY PRODUCTS OF THE NEUTRON AS WELL.*

FIGURE 7.5 Visualization of the proton (or neutron).

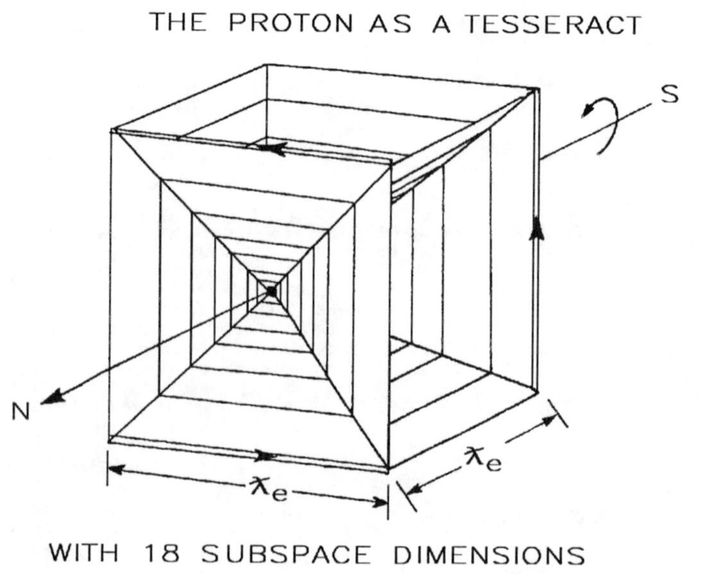

PROTON MAGNETIC MOMENT

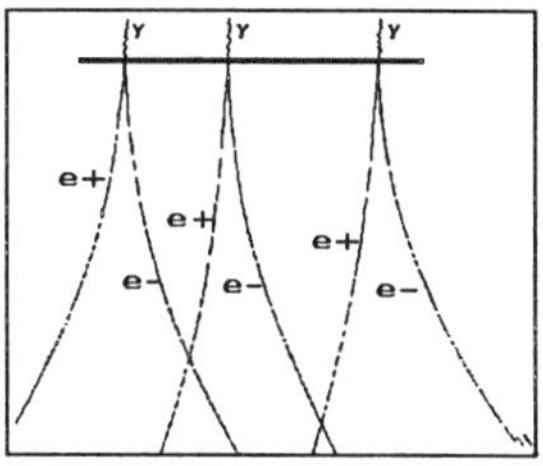

A magnetic moment is, by definition, a measure of the strength of a magnet. Magnetic moment relates to the turning effect (moment) in a magnetic field, and is defined as a torque *(t)* observed in a field at 90 degrees to the magnetic axis. The magnetic moment units are Ampere x meter squared or (J/T).

Subatomic particles are characterized by a specific mass and a specific magnetic moment. Magnetic moment requires a particle to have a mass, but there is not (necessarily) a direct relationship between a particle's mass and the value of a particle's magnetic moment. Composite particles were to surprise scientists, when it first became possible to measure their moments. It was universally assumed, at that time, that all subatomic particles would have a moment equal to one Bohr magneton. We now know that one Bohr magneton only applies to electrons and muons. *(FIGURE 8.0)* lists the expression for the Bohr magneton.

$$\mu_B = 1/2\left[e\,c\,\lambda_c\right] = \frac{e\hbar}{2\,m_x}$$

FIGURE 8.0 The Bohr magneton relationships.

Notice that mass (m_x) is in the denominator. The proton's magnetic moment actually measures about 2.79 Bohr magnetons. This denotes that not all of the proton's mass is actively magnetized. The presently accepted value, for the proton's magnetic moment, is measured by selectively (gyroscopically) resonating the small spinning core mass of the proton. This isolates the proton's magnetic moment from the much larger electron moments present in atoms.

EFFECTIVE MAGNETIC MOMENT:

In any resonant system, damping reduces the frequency of resonance to less than the natural frequency. Damping has been ignored, in all listings of the proton's magnetic moment, because the effect is inaccessible in the present experiments. We will show that there is a damping torque that is proportional to the polarizing measuring magnetic field, and the angle of tipping as the proton core precesses. Moreover, we will show that the angle of tipping is *QUANTIZED*, making the published value track the strength of the measuring magnetic field. The damping torque in magnetic systems can be expressed as; *(FIGURE 8.1)*

$$\textit{torque} \quad t = \mu_P \cdot \beta[\sin \emptyset]$$

$$\mu_P = \text{Joule/Tesla} \quad \beta = \text{Tesla}$$

FIGURE 8.1 Tipping angle torque between a magnetic moment in Joule per Tesla (J/T) and a magnetic field in Tesla (T)

The true value, for the proton's static magnetic moment, can be calculated using the same method as the electron in Chapter 4, but using the vector proton's core particle dimensions. The proton's circulating current is much larger than the electron model's current given in *(FIGURE 4.4)* . We can simplify the calculations by scaling directly from the electron's magnetic moment value taken from TABLE 1.0 APPENDIX A. *(see FIGURE 8.2)*

$$\mu_{pm} = \mu_e \left[\left[\frac{\sqrt{2}}{2}\right]^{18}\left[1-\sqrt{2}\,\alpha_\mu\right]\right]$$

Model

$$\mu_{pm} = 1.81045852 \times 10^{-26}\ \mathrm{A{\cdot}m^2}$$

CODATA

$$\mu_p = 1.41060761 \times 10^{-26}\ \mathrm{A{\cdot}m^2}$$

FIGURE 8.2 Scaling to the proton's true magnetic moment from the electron's magnetic moment measurement.

The calculated value is 1.81045852 x 10^{-26} J/T as shown. This is the true (undamped) proton magnetic moment, as taken from the model. Notice that the model's value is about 28 percent larger than the CODATA effective value listed. Because the model gives the total mass for the proton to within 0.000033 percent of the CODATA recommendation, we will be able to accurately deduce the damping correction and reconcile the true to the effective value for the proton's magnetic moment.

MEASUREMENT METHODS AND THEORY:

All methods to measure the proton's magnetic moment draw a parallel between a precessing top *IN A GRAVITY FIELD*, and the precessing proton core particle. But, these are two different systems. In a gravity field, a precessing spinning mass, for example, is not affected by the angle of tipping, as a consequence of the angular velocity being equal to the *RATIO* of two vector quantities acting through the same angle, so the sines cancel out. In a magnetic field, however, there is a torque that develops proportional to the magnetic moment, field strength, and tipping angle. The classical relationship, in a gravity field, for the independence of the angular velocity to the precession angle of tipping, is shown in *(FIGURE 8.3)*

$$2\pi f = \frac{MgR\,[\sin\phi]}{Mvr\,[\sin\phi]} = \frac{torque}{spin\ ang.\ mom.}$$

Where; M=mass
g=acc. gravity
R=radius arm
v=spin velocity
r=spin radius

FIGURE 8.3 Classical gravity gyroscope. Sine of the angle cancels making the angular velocity independent of the angle of precession.

The specious published value, for the proton's magnetic moment, is based on a parallel to the *(FIGURE 8.3)* ratio between the torque and spin angular momentum, as in *(FIGURE 8.4)*.

$$2\pi f = \frac{\mu_p \cdot \beta}{\frac{1}{2}\hbar}$$

FIGURE 8.4 This is used to determine the magnetic moment of the proton (speciously).

(FIGURE 8.4) is incomplete. In a magnetic field there is a direct reaction torque, to the angle of tipping. We will show that the angle of tipping is *QUANTIZED*. Being quantized, the damping error always maintains proportionality to the measuring parameters. The error proportionality tracks, making the damping error measurement undetectable. The true (undamped) proton magnetic moment is about 28 percent larger than the published value.

The use of timed 90 degree pulses in NMR measurements to enhance magnetic resonance signals, is not to be confused with the quantized *NUTATING* angle that results in the damping losses. *(FIGURE 8.5)* shows the constant nutating angle as the magnetic moment vector goes through large angles.

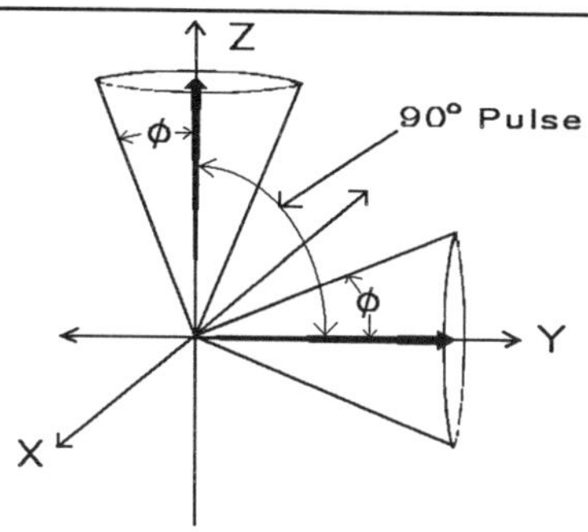

Magnetization vector changes during a 90° pulse. Nutation angle φ remains fixed.

FIGURE 8.5 Magnetization vector can change large angles but nutating angle is constant.

The proton's (predicted) true magnetic moment was deduced from the model's scaling and the electron's measured moment. *(FIGURE 8.2).* The model predicts, then, that the damping losses are exactly 3.9985091238 x 10^{-27} Joule per Tesla. We postulate that a reaction torque exists between the magnetic moment vector and the measuring field. The reaction torque (energy loss) from external driving sources, or loss from free induction decay (FID) can now be determined, using the model's geometry.

CORE PARTICLE TIPPING ANGLE:

The model's geometry is known, allowing the quantized precession angle (theta) to be determined, exactly. We visualize the core particle as being driven to the bottom wall by the strong measuring magnetic field, from whence the core particle tips, to alternately touch the left and right walls, of site#17. *(see FIGURE 8.6)*

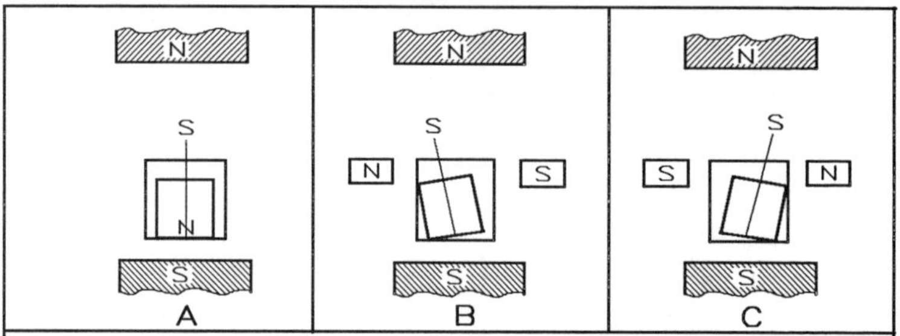

Model of proton's quantized tipping angle of precession.

FIGURE 8.6 Visualization of core particle tipping angle degrees of motion.

In *(FIGURE 8.6A)* the proton is shown to be closer to the south pole of the strong polarizing field (shaded pole pieces shown). The core is visualized as being driven to the bottom wall where it rocks back and forth in the alternating transverse (weak) excitation field, at the precession frequency. If the proton had happened to be closer to the north pole of the strong field, its core particle, site #18, would have been driven to the top wall of site #17. In both cases, the precession frequency would be the same. *(FIGURE 8.6B)* and *(8.6C)* demonstrate the fixed angle of tipping, in the proton model.

MODEL QUANTIZED TIPPING ANGLE:

We know the model's dimensions to extreme accuracy, based on the known fundamental physical constants, and the rules of geometry. The actual (site#18) anomalous magnetic moment protrusion, the separation between site #17 and #18, and the edge length of site #18 core particle are given in *(FIGURE 8.7A, B, and C)*.

(A) $$\alpha_{\mu 18} = \left[\frac{\sqrt{2}}{2}\alpha_\mu\right] \lambda_e \left[\left[\frac{\sqrt{2}}{2}\right]^{18} \left[1-\sqrt{2}\,\alpha_\mu\right]\right]$$

(B) $$d_{17} = \lambda_e \left[\left[\frac{\sqrt{2}}{2}\right]^{17} \left[1-\sqrt{2}\,\alpha_\mu\right]\right] \left[\frac{1-\left[\sqrt{2}/2\right]}{2}\right]$$

(C) $$L_{n18} = \lambda_e \left[\left[\frac{\sqrt{2}}{2}\right]^{18} \left[1-\sqrt{2}\,\alpha_\mu\right]\right]$$

FIGURE 8.7 (A) *the anomalous protrusion dimension*
(B) *separation between site#17 and #18*
(C) *edge length of core particle site#18*

Finally the theoretical tipping angle becomes *(FIGURE 8.8)*.

$$\sin \phi = \left[\frac{d_{17} - 2\alpha_{\mu 18}}{L_{n18} + \alpha_{\mu 18}}\right]$$

$$\sin \phi = 0.205299023341459$$

$$\phi = 11.84700414 \text{ DEGREES}$$

FIGURE 8.8 Calculation of proton model's tipping angle

The reader can verify that the algebra reduces *(FIGURE 8.8)* to the normalized *(FIGURE 8.9)*.

$$\sin \phi = \frac{\left[\left[\frac{1-\left[\sqrt{2}/2\right]}{2}\right] - \alpha_\mu\right]}{\left[\left[\sqrt{2}/2\right] + \left[\alpha_\mu/2\right]\right]}$$

FIGURE 8.9 Normalized tipping angle calculation.

PROTON MAGNETIC MOMENT PREDICTION:

Now that we have a theoretical value for the tipping angle, it is a relatively easy task to calculate the effect of the damping on the *GYROSCOPICALLY* measured magnetic moment.

The Bohr magneton can be transposed so that the spin

angular mometum times the fundamental charge will yield a mass when divided by a magnetic moment. Conversely, the spin angular momentum times the fundamental charge will yield a magnetic moment when divided by a mass. *(FIGURE 8.10)* We will use this relationship to calculate the gyroscopic magnetic moment with mass terms.

$$\text{Spin } \tfrac{1}{2}\hbar = m^2 \cdot kg \cdot s^{-1}$$
$$\text{Charge } e = A \cdot s$$
$$\tfrac{1}{2}\hbar\, e = [A \cdot \cancel{s}]\,[m^2 \cdot kg \cdot \cancel{s^{-1}}]$$
$$\tfrac{1}{2}\hbar\, e = m^2 \cdot kg \cdot A$$

FIGURE 8.10 Relationship between mass and magnetic moment from the Bohr.

We must consider the mass equivalent of the electrical potential energy at the core particle, site #18. The electrical charge dithers between both ends of the core particle, so we assume that half of the mass is effectively moved as the core tips. The mass from the Einsteinian then becomes: *(FIGURE 8.11)*

$$Mce = \left[\frac{e^2}{8\pi\, \varepsilon_0\, c^2\, [d18 - \alpha_\mu 18]}\right]$$

$$d18 = L_n 18 \left[\frac{1 - [\sqrt{2}/2]}{2}\right]$$

Where Mce is half electrical potential energy mass.
d18 is the separation to the vacuum of space at site #18

FIGURE 8.11 Half the mass equivalent of the electrical potential energy at the core particle.

Finally, the model's gyroscopic magnetic moment prediction is obtained starting with eqn. *(8.10)*. The spin angular momentum times the charge gives us a magnetic moment when divided by a mass. *(8.10)* and *(8.11)* mass estimates then combined for *(8.12)* magnetic moment.

(A) $$tpm = \left[\left[\frac{\frac{1}{2}\hbar e}{\left[\mu_{pm} - \left[\mu_{pm} \sin \phi\right]\right]}\right]\right]$$

(B) $$\mu_{pmg} = \left[\frac{\frac{1}{2}\hbar e}{\text{Mce} + tpm}\right]$$

$$\mu_{pmg} = 1.41065278 \times 10^{-26}\ \text{A}\cdot\text{m}^2$$

FIGURE 8.12 (A) Torque mass calculation
(B) Gyroscopic magnetic moment calculation, proton vector model.

The proton model's structure estimates the gyroscopic magnetic moment to within about 33 ppm of the CODATA recommendation.

SUMMARY:

The proton models as a composite particle whose core particle is not able to tip by more than a quantized tipping angle of 11.84700458 degrees. The magnetic moment vector reacts with the measuring magnetic field, as it wobbles, to produce a damping loss, by the well known *(FIGURE 8.1)*. As in any mechanical system, the frequency of precession will be reduced a measurable amount by the damping. The theory proposes that the damping effect has escaped notice because the error tracks the strength of the measuring magnetic field.

NEUTRON MAGNETIC MOMENT

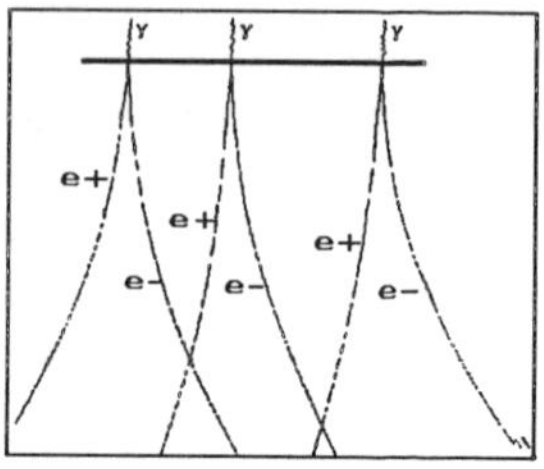

NEUTRON'S SURPRISE:

Magnetic moments require circulating charge currents around loop areas (Ampere times meter squared). The neutron, lacking a charge, was originally thought to have no magnetic moment. When the neutron was checked, by L. Alvarez and F. Block in 1940, it was found to have a magnetic moment of about (1.91) magnetons. The experiment used a resonance technique with a neutron beam obtained by deuteron bombardment of Be. Later (1949) Rodgers and Staub determined the sign of the moment to be negative as was suspected.

The neutron thus requires complex internal currents to have a magnetic moment and also have, at the same time, no apparent charge.

HELICITY OF MAGNETIC MOMENTS:

Helicity is the direction of the magnetic moment vector in relation to the spin. The electron is assigned negative helicity. The positron is assigned a positive helicity. Experimentally, the proton has a positive helicity and the neutron's magnetic moment has a negative helicity. The vector models support these experimental facts, as shown by an analysis of their structures. *(FIGURE 9.0).* Note that the charge conjugation due to the vector interactions of the electron type neutrino shifts the composite proton (first column) to a positive core, and the neutron (second column) to a negative core particle, for the proper helicity.

PROTON		NEUTRON	
0	+ Positron	0	∓ Electron / Positron
1	− Neutrino	1	∓ Neutrino / Neutrino
2	+ Neutrino	2	− Neutrino
3	− Neutrino	3	+ Neutrino
4	+ Neutrino	4	− Neutrino
5	− Neutrino	5	+ Neutrino
6	+ Neutrino	6	− Neutrino
7	− Neutrino	7	+ Neutrino
8	+ Neutrino	8	− Neutrino
9	− Neutrino	9	+ Neutrino
10	+ Neutrino	10	− Neutrino
11	− Neutrino	11	+ Neutrino
12	+ Neutrino	12	− Neutrino
13	− Neutrino	13	+ Neutrino
14	+ Neutrino	14	− Neutrino
15	− Neutrino	15	+ Neutrino
16	+ Neutrino	16	− Neutrino
17	− Neutrino	17	+ Neutrino
18	+ Neutrino	18	− Neutrino

NOTE: The PROTON starts with a positron and charge conjugation gives the core particle (18) a positive (+) magnetic moment. The NEUTRON starts with an electron (its decay product) plus an outer positron (site 0). At site (1) then, the neutrino acts negative, but the second neutrino acts positive shifting polarity of the neutron's core particle to be a negative magnetic moment (−) at site (18).

Proton and Neutron helicity from the charge conjugation

FIGURE 9.0 Polarity resulting from the charge conjugation between sites in the vector models of the proton (col. 1) and neutron (col. 2)

The neutron's mass and decay products were given, exactly, by postulating unformed pairs at the the outer site (O) and the next inner site (1) from FIGURE 7.4. We know, experimentally, that the neutron's magnetic moment exhibits a negative helicity. Accordingly, then, we can deduce how the neutron should stack up in the model. This is shown in the second column of *FIGURE 9.0.* The electron is one of the neutron's decay products, so an electron is shown as the outer most particle of the pair at site (O). (Presumably the antineutron would have a positron in that position.) There is no question about the stacking order of the electron type neutrino pair at the first inner site (1) because the electron type neutrino is its own antiparticle. (The same reasoning applies to the antineutron structure, as well.) As we follow the charge conjugation down the column layers we see that the polarity of the core particle site#18 is the correct negative helicity for the neutron's magnetic moment. (The antineutron will have a positive helicity by the same process.) As we shall see later, the charge currents short out at site (o) and the first site (1) then presents a positive polarity to site (2) so that overall charges balance out at nine positive and nine negative, giving the neutrino model its zero external charge.

Of course an antiproton would start with an electron at site (O) first column, from whence it would stack internally to give the correct negative helicity. Overall charge would be the correct negative for the antiproton, and charges would cancel out in the antineutron as required.

NEUTRON MAGNETIC MOMENT:

A clue as to the magnitude of the neutron's magnetic moment may be obtained from the hydrogen isotope (H_2). The deuteron is the nucleus and is composed of a proton and neutron in a bound state. The deuteron's measured magnetic moment is a vector result of the proton's and neutron's magnetic moment. From the known proton's positive magnetic moment, the neutron's apparent negative magnetic moment becomes; *(FIGURE 9.1)*.

Proton mag. mom.	$1.41060761 \times 10^{-26}$ J/T
Deuteron	$.43307375 \times 10^{-26}$ J/T
Neutron	$.97753386 \times 10^{-26}$ J/T
CODATA	$.96623707 \times 10^{-26}$ J/T

FIGURE 9.1 Deuteron's prediction for the neutron's magnetic moment, compared to the CODATA.

The slightly smaller CODATA value is theorized to be a scaler quantity so that the deuteron's magnetic moment is the result of a vector combination of the proton's positive magnetic moment and the neutron's negative magnetic moment. This explanation is the result of discovering that the deuteron has a quadrapole moment. A quadrapole moment implies that the magnetic moments of the deuteron's proton and neutron do not add directly.

The neutron model's magnetic moment calculation is complicated by the unformed pairs at site (O) and site

(1). What shift in charge current would result? We know, experimentally, the value of the neutron's magnetic moment, so we can try to deduce how the model might produce a similar value. We will assume that the published value for the neutron's magnetic moment is correct. We must also consider the theoretical "damping" at the core particle similar to that of the proton model. The magnetic moment, in the proton model, was obtained by simply scaling from the known electron's magnetic moment, to the proton's core particle. (CHAPTER 8). We could have, just as well, transferred the outer positrons circulating current to the smaller current loops of the proton's core particle, then corrected for damping. Before we use this method for the neutron's magnetic moment, it will be instructive to test the method on the proton's model *(FIGURE 9.2)*. First, the positron's circulating current *(the same value as the electron's circulating current given in FIGURE 4.4)* is calculated (A) and then scaled to the core particle's two Ampere turns (B).

(A) $$Ae = e \cdot c / \lambda_e = 19 \cdot 796339555 \text{ AMPS}$$

$$Ap = 2 \cdot Ae \left[\left[\frac{\sqrt{2}}{2} \right]^{18} \cdot (1 - \sqrt{2} \cdot \alpha_\mu) \right]^{-1}$$

(B) $$Ap = 2.030474066333 \times 10^{4} \text{ NA}$$

FIGURE 9.2 Proton core circulating Ampere turns.

Then the proton model's core edge length (L18) is scaled from the (electron's) positron's rationalized Compton wavelength ($\lambda\!\!\!^{-}_e$) as follows: *(FIGURE 9.3)*.

$$L18 = \lambda\!\!\!^{-}_e \cdot \left[\left[\frac{\sqrt{2}}{2} \right]^{18} (1 - \sqrt{2}\, \alpha_\mu) \right]$$

$$L18 = 7.5298091275 \times 10^{-16} (\text{m})$$

FIGURE 9.3 Core particle's edge length, meters.

The proton's loop area (LA) of the core particle becomes

the radius squared times pi. *(FIGURE 9.4).*

$$LA=\left[\left[\frac{L18}{2}\cdot\sqrt{2}\right]\cdot\left[1+\frac{\alpha\mu}{2}\right]\right]^2\cdot\pi \qquad (m^2)$$

$$LA=8.916432666730584\text{x}10^{-31}$$

FIGURE 9.4 Core particle's loop area, meters squared.

Finally, the proton model's true (undamped) magnetic moment (Upm) becomes: *(FIGURE 9.5).*

$$Upm=Ap\cdot LA=1.8104585294\text{x}10^{-26}\ J/T$$

FIGURE 9.5 Proton's true magnet moment calculated with transformed circulating currents.

This checks, as expected, with the 1.8104582 x 10^{-26} J/T true value proton magnetic moment that was obtained in *(FIGURE 8.2)* by simply scaling directly from the electron's experimentally measured magnetic moment. This exercise is taken as proof that the concept of circulating currents (in the composite vector models) may be extended to the neutron model as well.

NEUTRON MOMENT CALCULATIONS:

We postulate that the electron's circulating current is transformed to the first inner site (1) on account of the unformed pairs at the outer site (O). The circulating current (An) in the two core particle turns, site (18) is scaled through (18 minus 1) inner sites and becomes: *(FIGURE 9.6).*

$$An=2\cdot Ae\left[\left[\frac{\sqrt{2}}{2}\right]^{17}\cdot(1-\sqrt{2}\cdot\alpha\mu)\right]^{-1}$$

$$An=1.43576198132\ \text{x}10^{4}\ NA$$

FIGURE 9.6 Neutron circulating current transformed to the core particle Ampere turns.

Accordingly, then, the undamped neutron model's magnetic moment becomes from (9.4) and (9.6) as: *(FIGURE 9.7)*.

$$Unm = An \cdot LA = 1.28018750319 \times 10^{-26} \ J/T$$

FIGURE 9.7 Neutron's predicted true (undamped) magnetic moment.

We note that ($\sqrt{2}$) times the neutron's (undamped) magnetic moment *(9.7)* is exactly equal to the proton's (undamped) true magnetic moment *(9.5)* and *(8.3)*. If we postulate that the damping experienced by the proton may be extended to the neutron model by the same ($\sqrt{2}$) ratio, then the model would predict the neutron's gyroscopic magnetic moment as: *(FIGURE 9.8)*.

$$Upm = 1.8104585294 \times 10^{-26}$$
$$DPL = Upm - 1.41060761 \times 10^{-26}$$
$$Unm = 1.280187503196 \times 10^{-26}$$
$$Unmg = Unm - \left[\left[\frac{\sqrt{2}}{2}\right] DPL\right] \quad (J/T)$$
$$Unmg = 9.97450206624 \times 10^{-27}$$

FIGURE 9.8 Neutron model's best prediction for the damped magnetic moment.

Alternately, if we postulate that the electrical potential energy is the result of a reduced charge at the neutron's core particle, the neutron magnetic moment would be predicted as: *(FIGURE 9.9)*. It seems highly unlikely that the fundamental charge could be less than the value listed in Table 1, APPENDIX A, unless some new physics would allow it. Fractional fundamental charge can be argued against quite convincingly *(from FIGURE 1.1) making FIGURE 9.9 very unlikely.*

NOTE: (sin ϕ) see (8.9)
(Mce) see (8.11)

$$\tfrac{1}{2}\hbar e = 8.44806206397\text{x}10^{-54}$$

$$MKn = \frac{\tfrac{1}{2}\hbar e}{(Unm-(Unm \cdot \sin\phi))}$$

$$Unmg = \frac{\tfrac{1}{2}\hbar e}{(MKn + Mce)} \quad (J/T)$$

$$Unmg = 1.003225122896\text{7x}10^{-26}$$

FIGURE 9.9 Prediction using an unlikely reduced charge current to calculate damping losses for the neutron's magnetic moment.

SUMMARY AND CONCLUSIONS:

The neutron model gives the neutron mass to within 400 parts per billion of the CODATA, but the neutron is a complex structure that is difficult to develop into a theory for magnetic moments. We saw that the results of *(9.8)* or *(9.9)* are not precise, if one can believe the published values for the deuteron and the neutron's magnetic moments. I leave it to others to determine the source of the disagreement.

THE MUON'S STRUCTURE

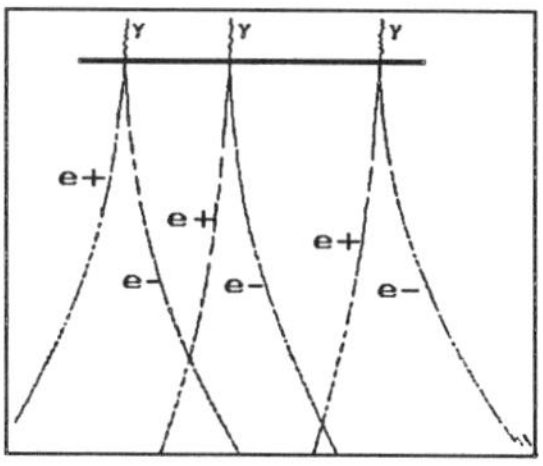

MUON TYPE NEUTRINOS:

In *FIGURE 3.3*, we created structures for the electron pair, and the single electron type neutrino which became its own antiparticle. We went on to construct models for the proton and neutron in Chapter 6. The remaining pair of particle models in *FIGURE 3.3*, were labeled the muon type neutrino pair. The muon neutrinos will now be used to construct the muon. We discussed the "ghostly" neutrinos and the specious claims for their detection in Chapter 5. We will now discuss the muon type neutrinos in more detail.

It was thought that the muon's decay, which produces two neutrinos, should have a cross section to have the neutrinos annihilate into photons. Neutrinos are leptons, and leptons are always produced in conjugate pairs. The long sought reaction *(FIGURE 10.1)* has never been detected. The reason for this null is *NOT* that the muon decay neutrinos are a *MIXED SPECIES*, but rather, as this model indicates, *FREE NEUTRINOS CANNOT MOVE*. Annihilation into photons requires positioning E on one with B on the other. By not moving, the neutrinos cannot present conjugate structural members, to their mates, thus preventing annihilation.

$$\mu\pm \longrightarrow e\pm + \nu_\mu + \bar{\nu}_\mu$$

$$\mu\pm \not\longrightarrow e\pm + \gamma$$

NEUTRINOS CAN'T MOVE, THEREFORE, THEIR CONJUGATE MEMBERS DO NOT MEET TO CAUSE THE ANNIHILATION INTO PHOTON(S) (γ)

FIGURE 10.1 Annihilation of muon neutrino pairs does not happen because free neutrinos can't move to present conjugates to their mates.

MUON CHARACTERISTICS:

The muon, experimentally, decays into an electron (or positron) and two neutrinos. The muon has an average lifetime of approximately 2.2 x 10^{-6} seconds. This decay time is a relatively long time compared to the lifetimes of other mesons. The muon is actually the remnants of the pi meson decay, but we will not attempt to model the pi meson. Despite the short life of the muon, the muon can be studied by giving it a relativistic velocity and producing time dilation. CERN, in Europe, measures the characteristics of the muon, in a specially designed synchrotron apparatus, to at least nine decimal places, with time dilation factors of on the order of 30 times.

The muon acts, in all respects, as a miniature electron (or positron) rather than the composite particle that it is. For example, the muon's magnetic moment is nearly equal to one Bohr magneton, just like the basic electron. No other composite particle has a magnetic moment that involves all of its mass. The muon has a mass equal to about 206.76 electron masses and is a composite particle.

The muon is the champion penetrating particle. Cosmic ray muons, produced in the earths atmosphere, are routinely detected deep underground. No other charged particle, regardless of its mass, can penetrate more than a few meters in rock. The use of the terms "weak" or "strong" force is a superficial attempt to explain the difference between heavy hadrons and the medium weight muon's penetrating ability, without an actual mechanism. These models show a physically smaller cross section for the muon. A physically smaller size provides an explanation for the muon's penetrating ability. We will show that the muon is actually a collapsed electron or positron. This model, then, explains the one Bohr magneton muon, and explains the penetrating ability without having to resort to abstract terms, that have no visual basis.

MUON MODEL:

The muon type neutrinos act in pairs with the electron (or positron) to form the muon. As with the electron type neutrinos, the muon neutrinos absorb energy, become smaller and form inside of the electron's (positron's) vortex. We combine vectors, member by member, as in the scheme used to form the excellent proton and neutron models. (CHAPTER 6). The (FIGURE 10.2A and 10.2B) demonstrate these models parallel nature by combining to form the muon. In (10.2A), the electron first reacts with the muon neutrino to form an exotic ($) structure. Notice there is one normal rotating face with the remaining virtual cube structure made up of only (E) type vectors. This exotic vortex then reacts with a muon type antineutrino, vector by vector, to create a virtual electron type structure (the negative muon model). We postulate that the exotic ($) causes the vectors to crowd to one site, making the muon act *(LIKE IT DOES EXPERIMENTALLY)* as a miniature electron. The reader can review *(FIGURE 3.3)* and verify that the muon type neutrinos, vector by vector, do indeed virtually reproduce the electron's structure in miniature.

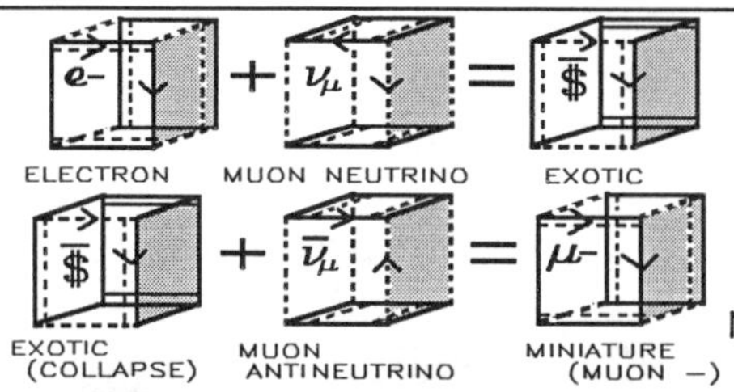

FIGURE 10.2 A

ELECTRON'S COLLAPSE TO FORM THE NEGATIVE MUON

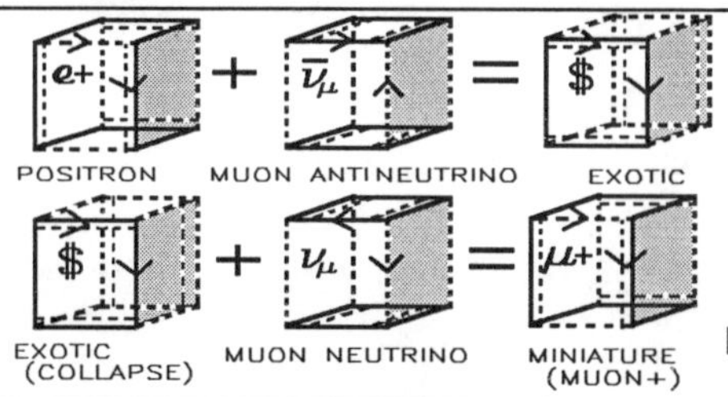

FIGURE 10.2 B

POSITRON'S COLLAPSE TO FORM THE POSITIVE MUON

Similarly, (10.2B) shows the action with the positron to produce the positive muon. Here we start with the muon type antineutrino to produce an exotic ($) with one normal rotating face and the remaining cube structure entirely made up of magnetic (B) vectors. This exotic reacts to form, vector by vector, a virtual positron type structure (the positive muon model).

MUON MODEL SPIN AND MOMENT:

The muon's fundamental physical constants can be used to adjust the size of the muon model's cube, in the same fashion as the electron (CHAPTER 4). *AS THIS MODEL SHOWS, THE MUON IS A COLLAPSED MINIATURIZED ELECTRON.* The muon's magnetic moment, theoretically, should be equal to one Bohr magneton, as calculated in *(FIGURE 10.3).* Much as the electron, the muon's magnetic moment measures slightly larger than the Bohr. CERN measures the muon's magnetic moment to about nine decimal places in their synchrotron. The muon's anomalous magnetic moment gives a 'g' factor that is larger than the electron's 'g' factor. *(FIGURE 10.4).*

$$\mu_{\mu B} = \tfrac{1}{2}(e\, c\, \lambda\!\!\!^{-}_{\mu}) = 4.4852219 \times 10^{-26}\,\mathrm{J/T}$$

WHERE: $\mu_{\mu B} =$ MUON BOHR MAGNETON

$\lambda\!\!\!^{-}_{\mu} = 1.86759475 \times 10^{-15}\,\mathrm{m}$ MUON RATIONALIZED COMPTON WAVELENGTH

FIGURE 10.3 The muon's Bohr magneton calculation

$$\alpha_{\mu} = 2((1.00116593994)^{1/2} - 1)$$

$$\alpha_{\mu} = .00116560029$$

$$\mu_{\mu} = \tfrac{1}{2}(e\, c\, \lambda\!\!\!^{-}_{\mu}) \times (1 + \alpha_{\mu}/2)^2$$

$$\mu_{\mu} = 4.49045140 \times 10^{-26}\,\mathrm{J/T}$$

FIGURE 10.4 The muon's anomalous magnetic moment.

This yields, as expected, the CODATA value for the

muon's magnetic moment from the published (μ_μ). The collapsed electron model would suggest that the larger 'g' factor is the result of the extra vectors crowding to one site of a single cube. The spin angular momentum is considered to equal ($\frac{1}{2}\hbar$). Again the mass radius of gyration is equal to half of the muon's rationalized Compton wavelength. The reader can verify that the equation *(FIGURE 10.5)* is correct numerically.

$$\tfrac{1}{2}\hbar = Rm \times m_\mu \times c$$
$$= 5.272863345\times10^{-35} \text{ J.s}$$

WHERE:

$Rm = \tfrac{1}{2}\bar{\lambda}_\mu$ mass radius (muon)

$m_\mu =$ muon mass

$c =$ velocity of light

FIGURE 10.5 Check of spin angular momentum from the muon model's dimensions of the spinning cube structure.

SUMMARY:

We conclude that the muon is a collapsed electron (or positron) based on the coherence of these vector models, and the experimental evidence that it is so. The muon mass is not easily predicted by this model, but the muon is formed when the pion decays. This suggests that the pi meson is a defective proton (or antiproton) formed with muon type rather than electron type neutrinos. The electron type neutrinos produce charge conjugation and it is the resulting electrostatic attraction, between layers in the proton, that gives the proton a stable structure. But note that the muon type neutrinos *DO NOT CAUSE CHARGE CONJUGATION*, consequently there are no electrostatic forces to hold the pi meson together. The muon, then, is probably the core of the pion, and the muon lasts longer because its vectors are crowded to one site along with the collapsed electron.

In this view, as experiments show, all mesons are composed of the muon type neutrinos, so they all will quickly decay, and play no role in the scheme of things.

THE FINE STRUCTURE CONSTANT CALCULATIONS

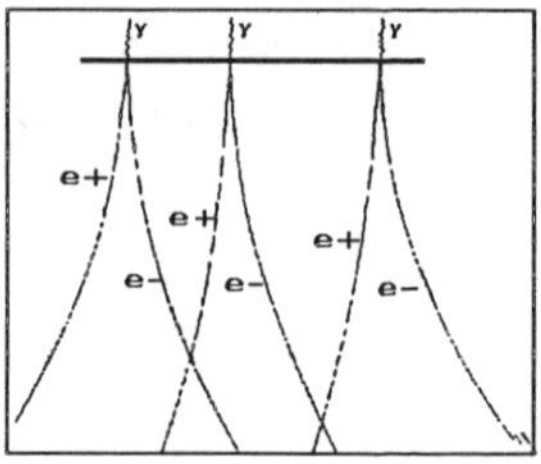

We used the fine structure constant in calculating the electron's anomalous magnetic moment, *FIGURE 1.2.* Originally, the fine structure constant was used to explain splitting of different levels in atomic spectra, and it is from this that it got its name. Why the fine structure constant (α) has the value it does, is a mystery. Theoretical physicists, ever since it was discovered 50 years ago, have studied the fine structure constant and wondered about it.

The fine structure constant (α) is a dimensionless *RATIO* that relates the interaction between a charged particle and the electromagnetic field.

There are a number of ways one might derive the fine structure constant. A review of some methods would seem to be the best way to find a clue as to its cause. As we have seen, with this vector model, *ALL PRECISION SCALES FROM THE EXACT PLANCK SIZE OF THE ELECTRON.* The fine structure constant is no exception, its value can be obtained by dimensional analysis and the metric of the fundamental physical constants for the photon and electron. To calculate the fine structure constant, then, we will dissect the derived metric units into their basic units of meter (m) kilogram (kg) second (s) and Ampere (A). It is convenient to reduce all such compound (derived) units to an equivalent length (m) to obtain the dimensionless ratio for (α).

First, the dimension of the permeability of vacuum (μ_o) *FIGURE 11.0.*

$$\mu_0 = 4\pi \times 10^{-7}\ (\mathrm{m \cdot kg \cdot s^{-2} \cdot A^{-2}})$$

FIGURE 11.0 Permeability of vacuum.

To remove all dimensions except the length (m) by inspection, we multiply by the electron's charge squared or ($e^2 = s^2 \cdot A^2$) and divide by the electron's mass (m_e) in kilograms (kg). The calculations, using the electron's fundamental physical constants yields: *FIGURE 11.1*

$$\mathrm{m1} = \frac{\mu_0 \cdot e^2}{m_e} = 3.541128994 \times 10^{-14}\ \mathrm{meter}$$

FIGURE 11.1 Photon's electric component length.

The dimensionless fine structure constant (α) may now be obtained by dividing the meters of *(11.1)* by the electron's Compton wavelength (λ_e) in meters. *FIGURE 11.2*

$$\alpha = \frac{\mathrm{m1}}{2\lambda_e} = \frac{1}{137.03598} = 7.297353074 \times 10^{-3}$$

FIGURE 11.2 Derived fine structure constant as a ratio between the photon's electric length and the electron's Compton wavelength.

By a similar process we can retrieve a length from the permittivity of vacuum (ϵ_o) *FIGURE 11.3*

$$\epsilon_o = 8.854187817 \times 10^{-12}\ (\mathrm{m^{-3} \cdot kg^{-1} \cdot s^4 \cdot A^2})$$

FIGURE 11.3 Permittivity of vacuum.

Again, by inspection, we can retrieve a length (m) by multiplying (ϵ_o) by the velocity of light squared ($c^2 = m^2 \cdot s^{-2}$) the electron's mass (m_e) in (kg) and, to change the signs of the remaining $(m^{-1} \cdot s^2 \cdot A^2)$ we divide into the charge squared ($e^2 = s^2 \cdot A^2$). *FIGURE 11.4*

$$m2 = \frac{e^2}{m_e \cdot c^2 \cdot \epsilon_0} = 3.541128994 \text{x} 10^{-14} \text{ meter}$$

FIGURE 11.4 Photon's magnetic component length.

It is interesting to note that (m1) and (m2) are identical, and either will obtain the fine structure constant using the electron's Compton wavelength, in ratio.

THE VECTOR ELECTRON:

We can deduce the fine structure constant from the vector electron's prediction of two current loops. We treat the two current loops as if they were a two turn electromagnet. In electrical engineering there is an expression for the magnetic field in Tesla (T) of a coil as related to the permeability (μ_o) times the Ampere (A) turns (N) per unit length (m). *FIGURE 11.5*

$$T = \frac{\mu_0 \, NA}{m} = 8.828011157 \text{x} 10^9$$

Where:

$$T = \frac{\text{Einsteinian (J)}}{\text{Bohr (J/T)}}$$

$$A = e \cdot c / \lambda_e = 19.79633951$$

$$N = 2$$

$$\mu_0 = 4\pi \text{ x } 10^{-7}$$

FIGURE 11.5 Vector electron's magnetic field strength, in Weber per square meter (T).

The vector electron is smaller than any atom so that the permeability (μ_o) was taken as that of vacuum (4π) x 10^{-7} H/m. We obtained the electron's ideal magnetic field strength in *(11.5)* from the Bohr (J/T) where (J) is the electron's Einsteinian ($m_e \cdot c^2$). It follows, from the algebra, that the expression for the electron model's magnetic coil length (m3) becomes: *FIGURE 11.6*

$$m3 = \frac{\mu_0 \cdot 2A}{T} = 5.635881846 \times 10^{-15} \text{ meter}$$

FIGURE 11.6 Electron model's magnetic coil length.

The fine structure constant then becomes, where (λ_e) is the electron's rationalized Compton wavelength: *FIGURE 11.7*

$$\alpha = \frac{m3}{2\lambda_e} = \frac{1}{137.03598} = 7.297353074 \times 10^{-3}$$

FIGURE 11.7 The fine structure ratio from the vector electron's prediction of a two turn magnetic coil in the electron.

It would appear that the coil length (m3) is a radius vector of the circumferential vacuum photon lengths. *FIGURE 11.8*

$$m1 = m2 = (m3 \times 2\pi)$$

CHECK:

$$\frac{3.541128994 \times 10^{-14}}{5.635881846 \times 10^{-15}} = 2\pi$$

FIGURE 11.8 Electron coil length as a radius vector.

It appears that the fine structure constant, like all physical processes, is geometric in origin, and derives its precision from the electron as a quantum sized chunk of mass (energy).

APPENDIX A

The international metric system of units is used exclusively in this monogram. The SI (System International) has been adopted world wide and is the modern language of science.

The SI is the current version of the familiar MKSA system of units (Meter, Kilogram, Second, Ampere) and has a simple hierarchy of relationships. As such, the SI serves to reduce errors in designs by being coherent, preventing misplaced decimal points, and being easily interpreted.

BASIC UNITS:

METER: The SI unit of length is the meter. The length of a meter was originally based on a measure of the earth's quadrant. A surveyors estimate of length produced a brass meter rod (artifact) that is now obsolete as a standard of length. A 1975 international agreement to fix the velocity of light at 2.99792458×10^8 meters per second, makes it possible to determine the length of a meter, by an accurate measurement of time in seconds. The meter is represented by (m).

KILOGRAM: The SI unit of mass is the kilogram, and is equal to the international prototype (artifact) mass. The original kilogram was based, historically, on the mass of one liter of water under standard temperature conditions. The kilogram is the only basic unit represented by a prefix "kilo" which means 1000 grams. The gram was too small a unit to be used with the meter in a coherent set of units. The kilogram symbol is (kg).

SECOND: The SI unit for time is the second. The modern second is standardized as the duration of

9 192 631 770 periods of the radiation corresponding to the transition between the two hyperfine levels of the ground state of the caesium-133 atom. This atomic clock measures time accurately to the 12th decimal place. The extreme accuracy of time allows for an accurate length of the meter by fixing the velocity of light as exact. The symbol for time is (s).

AMPERE: The SI unit for electric current is the Ampere. By definition the Ampere is that current which, if maintained in two straight conductors, of infinite length, placed one meter apart, would produce between those conductors, a repelling force of 2×10^{-7} Newton per meter of length. In practical measurements the Ampere is determined by using a current balance, rather than using the difficult, if not impossible conditions specified in the definition. The symbol for the Ampere is (A).

DERIVED UNITS:

The basic meter (m), kilogram (kg), second (s) and Ampere (A) are used to form derived standard units made up of products of the basic units. Those of interest to the fundamental constants discussed in this monogram are given in *FIGURE A1*.

(1 A) Force, Newton (N) = $(kg \cdot m \cdot s^{-2})$

(1 B) Energy, Joule (J) = $(N \cdot m) = (m^2 \cdot kg \cdot s^{-2})$

(1 C) Capacitance
Farad (F) = $(m^{-2} \cdot kg^{-1} \cdot s^4 \cdot A^2)$

(1 D) Charge, Coulomb (C) = $(A \cdot s)$

(1 E) Magnetic Flux
Weber (Wb) = $(m^2 \cdot kg \cdot s^{-2} \cdot A^{-1})$

(1 F) Magnetic Flux Density, Tesla (T)
(T) = $(Wb/m^2) = (kg \cdot s^{-2} \cdot A^{-1})$

(1 G) Magnetic Moment (J/T) = $(A \cdot m^2)$

FIGURE A1 Selected derived units of measure used in this vector particle model.

In *FIGURE A1*, equation (1A), The force is mass (kg) times acceleration in meters (m) per second per second (s^{-2}). Equation (1B) is the force (N) acting over a length (m) giving energy (J) in Joule. *(Note: Torque is always listed in* (N.m) *being a vector quantity, as opposed to energy which is a scaler. Both, however, have the same basic units.)* Equation (1C) is the unit of capacitance (Farad) and has the dimensions of the charge squared (C^2) divided by the energy (J). Equation (1E) the magnetic flux is (Weber) and has the dimensions of (J/A) and equation (1F) for magnetic flux density (Tesla) becomes Weber per square meter. The magnetic moment (1G) is a measure of the energy (J) per Tesla (T), which reduces to simply ($A.m^2$).

FUNDAMENTAL PHYSICAL CONSTANTS:

The fundamental physical constants, listed in TABLE 1, are known to parts per million precision. These constants are, in general, related to each other by a number of exact mathematical relationships. The mathematics allows (NIST) to assure that the list of constants have self consistent values, using only a small subset of constants for which they have accurate data. The resulting column "Correlated (relative) Uncertainties" are listed in parts per million (ppm). It is one of the great unsolved mysteries of science as to why the constants remain in such precise RATIOS, regardless of the system of units one might choose to measure them with. The vector model's predictive powers reflect the underlying geometry that makes it so. We will model the basic particles and demonstrate that, for our limited look, geometry serves quite well to scale the nature of subatomic particles that last longer than 2.2 microseconds (10^{-6} s).

Note that TABLE 1 lists the speed of light (c), the permeability of vacuum (μ_o) and the permittivity of vacuum (ϵ_o) as exact. They are exact by international agreement. In 1975, the speed of light was fixed at

2.99792458 x 10^8 m/s. Since (μ_o) is (historically) set at exactly (4π) x 10^{-7} Henry per meter, (ϵ_o) in Farad per meter is also known exactly from *FIGURE A2.*

(A) $$c = \frac{1}{\sqrt{\epsilon_o \mu_o}}$$

$c = 2.99792458 \times 10^8$
$\mu_0 = 4\pi \times 10^{-7}$ H/m
$\epsilon_0 = 1/(c^2 \cdot \mu_0)$ F/m

$\epsilon_0 = 8.8541878176 \times 10^{-12}$
$(m^{-3} \cdot kg^{-1} \cdot s^4 \cdot A^{+2})$

FIGURE A2 Exact calculation of the permittivity of the vacuum of space from the exact (by definition) velocity of light (c).

Equation (A) is the well known Maxwell "ether" resonance equation for the speed of light. Today, of course, we no longer believe in a conducting "ether" medium, rather we believe that the photon travels at (c) by a mutual embrace within its own structure. The photon literally makes its own space as it defines space and time.

All is not rosy in the TABLE 1 list of constants. Some of the physical constants, notably mass and magnetic moment, cannot be (directly) related. Physics has had great fun trying to come up with ways to do that, with varying degrees of success. The here proposed vector models are a step in that direction.

TABLE 1
Fundamental Physical Constants (1986 CODATA)

Quantity	Symbol	Value	Units	Correlated (relative) Uncertantity (ppm)
speed of light in vacuum	c	299 792 458	$\mathrm{m\,s^{-1}}$	(exact)
permeability of vacuum	μ_o	$4\pi \times 10^{-7}$	$\mathrm{N\,A^{-2}}$	
		$=12.566\,370\,614\ldots$	$10^{-7}\,\mathrm{N\,A^{-2}}$	(exact)
permittivity of vacuum	ϵ_o	$1/\mu_o c^2$		
		$=8.854\,187\,817\ldots$	$10^{-12}\,\mathrm{F\,m^{-1}}$	(exact)
Newtonian constant of gravitation	G	6.672 59(85)	$10^{-11}\,\mathrm{m^3\,kg^{-1}\,s^{-2}}$	128
Planck constant	h	6.626 0755(40)	$10^{-34}\,\mathrm{J\,s}$	0.60
$h/2\pi$	$\hbar$	1.054 572 66(63)	$10^{-34}\,\mathrm{J\,s}$	0.60
elementary charge	e	1.602 177 33(49)	$10^{-19}\,\mathrm{C}$	0.30
magnetic flux quantum, $h/2e$	Φ_o	2.067 834 61(61)	$10^{-15}\,\mathrm{Wb}$	0.30
Josephson frequency–voltage ratio	$2e/h$	4.835 9767(14)	$10^{14}\,\mathrm{Hz\,V^{-1}}$	0.30
electron mass	m_e	9.109 3897(54)	$10^{-31}\,\mathrm{kg}$	0.59
proton mass	m_p	1.672 6231(10)	$10^{-27}\,\mathrm{kg}$	0.59
proton-electron mass ratio	m_p/m_e	1836.152 701(37)		0.020
fine-structure constant, $\mu_o ce^2/2h$	α	7.297 353 08(33)	10^{-3}	0.045
inverse fine-structure constant	α^{-1}	137.035 9895(61)		0.045

Quantity	Symbol	Value	Units	Relative uncertainty (ppm)
proton magnetic moment	μ_p	1.410 607 61(47)	10^{-26} J T^{-1}	0.34
in Bohr magnetons	μ_p/μ_B	1.521 032 202(15)	10^{-3}	0.010
in nuclear magnetons	μ_p/μ_N	2.792 847 386(63)		0.023
proton gyromagnetic ratio	γ_p	26 752.2128(81)	10^4 s^{-1} T^{-1}	0.30
	$\gamma_p/2\pi$	42.577 469(13)	MHz T^{-1}	0.30
neutron mass	m_n	1.674 9286(10)	10^{-27} kg	0.59
neutron–electron mass ratio	m_n/m_e	1838.683 662(40)		
neutron magnetic moment	μ_n	0.966 237 07(40)	10^{-26} J T^{-1}	0.41
in Bohr magnetons	μ_n/μ_B	1.041 875 63(25)	10^{-3}	0.24
muon mass	m_μ	1.883 5327(11)	10^{-28} kg	0.61
muon–electron mass ratio	m_μ/m_e	206.768 262(30)		
muon magnetic moment	μ_μ	4.490 4514(15)	10^{-26} J T^{-1}	0.33
in Bohr magnetons,	μ_μ/μ_B	4.841 970 97(71)	10^{-3}	0.15
muon magnetic moment anomaly $[\mu_\mu/(e\hbar/2m_\mu)] - 1$	a_μ	1.165 9230(84)	10^{-3}	7.2
muon g-factor, $2(1 + a_\mu)$	g_μ	2.002 331 846(17)		0.0084

Quantity	Symbol	Value	Units	Relative uncertainty (ppm)
electron				
Compton wavelength, $h/m_e c$	λ_e	2.426 310 58(22)	10^{-12} m	0.089
$\lambda_e/2\pi = \alpha a_o = \alpha^2/4\pi R_\infty$	$\bar{\lambda}_e$	3.86 159 323(35)	10^{-13} m	0.089
electron magnetic moment	μ_e	928.477 01(31)	10^{-26} J T^{-1}	0.34
in Bohr magnetons	μ_e/μ_B	1.001 159 653 193(10)		1×10^{-5}
in nuclear magnetons	μ_e/μ_N	1838.282 000(37)		0.020
electron magnetic moment anomaly, $\mu_e/\mu_B - 1$	a_e	1.159 653 193(10)	10^{-3}	0.0086
electron g-factor, $2(1+a_e)$	g_e	2.002 319 304 386(20)		1×10^{-5}

NOTE: CODATA is the Committee on Data for Science and Technology who publish the recommended values for the constants based on data collected by the (USA) National Institute for Standards and Technology (NIST) (Taylor and Cohen). A complete summary of CODATA is available in Bulletin 63, November 1986 Pergamom Journals Inc. Maxwell House, Fairview Park, Elmsford, NY 10523

APPENDIX B

COMPUTER MODELING PROGRAM DETAILS:

Refer to the attached BASIC program listing. The program may be summarized as follows, by line numbers: (Read also the statements given throughout the program listing.)

Lines 10-70, first we declare the metric constants as exact for the calculations. (See also APPENDIX A)

Lines 80-180, produces a ratio of mass to length as a derived constant of nature.

Lines 190-310, calculates the alternating power series, and the anomalous magnetic moment protrusions. (See FIGURE 1.2 and FIGURE 4.5 for details.)

Lines 320-1020, provides a listing of the constants chosen for program operations.

Line 1030, sets the dimension of the arrays in memory.

Lines 1040-1070, resets variable names.

Line 1080, normalized separation. (See FIGURE 6.8 for details)

Line 1090, electrical potential energy effective mass ratio. (See FIGURE 6.3 for details.)

Line 1100, normalized protrusion factor. (See FIGURE 6.6 and FIGURE 6.7 for details.)

Line 1120, unscaled, normalized mass ratio of 1 (positron to electron mass ratio) divided by the protrusion factor.

Lines 1130, 1140, establishes the first values of the memory arrays reserved on line 1030.

Lines 1160-1200, this loop calculates the scaling for each inner neutrino and stores the results in the array

Lines 1340-1360, resets T, T1, T2, totals to zero.

Lines 1370-1510, this loop retrieves V(Z), V(Z+1), S(Z), S(Z+1) in turn from the memory array and keeps a running total as T, T1, T2. D3 and D4 are totals of the effective mass ratios of the photon structure plus the electrical potential energy.

Line 1520, calculates the total electrical potential energy between sites in millions of electron volts, MeV.
Line 1580, calculates neutron mass ratio.
Lines 1590-1750, sends results to printer and compares totals to the CODATA values,and calculates deltas.
Lines 1760-1820, calculates and sends to printer the proton model's true magnetic moment. (See FIGURE 4.5 and FIGURE 8.2 for details).
Lines 1840-2090, calculates and sends to the printer the proton model's prediction for the gyroscopic magnetic moment. (See FIGURE 8.7, 8.9, 8.10, 8.11 and 8.12 for details.)
Lines 2150-3480, calculates the model and prints to screen.
Lines 3530-3570 allows choosing ANY mass for the electron and still maintain the proper ratio between a (fictitious) mass and length. If one uses the CODATA electron mass, of course, we get the electron's rationalized Compton wavelength length as (L1).
Lines 3580-3740, this subroutine calculates values for the (fictitious) physical constants using a (fictitious) electron. (See FIGURE 1.0 for details.) If we had used the CODATA electron mass (Line 3550) then this subroutine would calculate a good set of CODATA fundamental physical constants. (An interesting test).
Lines 3750-4090, this subroutine allows us to calculate the models with either NONSENSE or CORRECT sets of fundamental physical constants. This allows a recheck of the constants calculated in Lines 3580-3740. The set must be consistent to work.

NOTE: The subroutine algorithm (Line 3580-3740) always returns a consistent set of constants. If you entered a nonsense value for the electron's mass (Line 3550) the proton's dimensionless mass ratio would be calculated about the same value. This demonstrates that the proton model's structure is maintaining the proper ratios between the several constants only if they are a consistent set. (The magnetic moments may not be correct, because they are not dimensionless ratios.)

BASIC COMPUTER PROGRAM (IBM PC)

This program calculates the structures for the proton and neutron, and the proton's magnetic moment. The program is in double precision for the IBM PC with Microsoft (TM) GWBASIC/D. The program may also run on other computers set up for BASIC. If you key this program take care to add the spaces as shown so that the tables format correctly.

```
***********************************************************************

10 CLS 'FIRST WE ESTABLISH THE SET OF EXACT VARIABLES
20 C#=299792458#  'METERS PER SECOND
30 A8#=.00729735308#  'DIMENSIONLESS CONSTANT
40 H1#= 1.05457266D-34  'JOULE x SECONDS
50 PI#=3.1415926536#  'DIMENSIONLESS CONSTANT OF PI
60 U#= (4*PI#)*.0000001#  'HENRY PER METER PERMEABILITY OF SPACE
70 G#= 1/(((C#)^2)*U#)  'FARAD PER METER PERMITTIVITY OF SPACE CALCULATION
80 PRINT "NOTE: WITH THE EMPERICAL VALUES FOR THE VELOCITY OF LIGHT, THE"
90 PRINT "FINE STRUCTURE CONSTANT AND THE PLANCK CONSTANT, IT IS
POSSIBLE"
100 PRINT "TO DETERMINE OTHER RELATED CONSTANTS NEEDED FOR THE MODEL."
110 PRINT "**********"
120 PRINT "WE ESTABLISH A NEW CONSTANT (KM#) EQUAL TO THE RATIO OF"
```

```
130 PRINT "THE CODATA 1986 VALUES FOR THE ELECTRON'S MASS AND ITS"
140 PRINT "RATIONALIZED COMPTON WAVELENGTH.  THIS ASSURES THAT"
150 PRINT "CONSISTENT SETS OF CONSTANTS WILL PRODUCE IDENTICAL"
160 PRINT "ELECTRICAL POTENTIAL ENERGY (MASS EQUIVALENTS)."
170 KM#=2.358971849D-18   'KILOGRAMS PER METER, DERIVED CONSTANT.
180 '****
190 'FOR LATER USE, THE RATIO (FINE STRUCTURE/PI) IS CALCULATED.
200 AA#=A8#/PI#   'DIMENSIONLESS
210 PRINT "NOTE:*********"
220 PRINT "WE OPT TO DETERMINE THE VALUE FOR THE ELECTRON'S 'g'"
230 PRINT "FACTOR FROM A (WITHOUT QED) ALTERNATING POWER SERIES"
240 PRINT "OF THE FINE STRUCTURE CONSTANT DIVIDED BY PI FOR"
250 PRINT "SUBSEQUENT CALCULATION OF THE PROTON MODEL'S SCALING,"
260 PRINT "RATHER THAN USE THE CODATA PUBLISHED VALUE FOR 'g'."
270 AE#=((1/2)*(AA#))-((1/3)*((AA#)^2))+((1/4)*((AA#)^3))-((1/5)*((AA#)^4))
+((1/6)*((AA#)^5))  'DIMENSIONLESS ALTERNATING POWER SERIES CALCULATION
280 'NOTE:***
290 'THE MODEL'S 'g' FACTOR AS A NORMALIZED RADIUS EXTENSION'
300 '(AU#) FOR ANOMOLOUS MAGNETIC MOMENT AND SCALING CALCULATIONS.'
310 AU#=2*((SQR(1+AE#))-1)
320 PRINT "***********"
330 PRINT"PLEASE NOTE: YOU SHOULD HAVE STARTED WITH  GWBASIC/D "
340 PRINT"SO THAT DOUBLE PRECISION RESULTS ARE OBTAINED FOR SQUARE
```

```
ROOTS"
350 PRINT"OR PROGRAM WILL NOT WORK AS CLAIMED!!!!!!!!!!!!!"
360 PRINT
370 PRINT"PRESS AND ENTER ANY KEY TO CONTINUE"
380 INPUT X$
390 CLS
400 PRINT "DO YOU WISH TO HAVE THE CONSTANTS CALCULATED?"
410 PRINT "TYPE 'Y' FOR YES, OR 'N' FOR NO"
420 INPUT B$
430 IF B$="Y" OR B$="y" THEN GOSUB 3520
440 IF B$="Y" OR B$="y" THEN 600
450 IF B$="N" OR B$="n" THEN 470
460 GOTO 420 'LOOP FOR CHOICES IN UPPER OR LOWER CASE
470 CLS
480 PRINT "DO YOU WANT TO SPECIFY THE PHYSICAL CONSTANTS?"
490 PRINT "TYPE 'Y' FOR YES  OR 'N' FOR NO:  NOTE: WHEN YOU SAY YES (Y)"
500 PRINT"TRY BOTH A CONSISTENT SET AND AN INCONSISTENT SET OF
CONSTANTS."
510 PRINT"THIS WILL SHOW THAT THE CALCULATIONS REQUIRE CONSISTENT SETS
TO"
520 PRINT"YIELD ESSENTIALLY THE SAME DIMENSIONLESS MASS RATIOS!!!!!"
530 INPUT R$
540 CLS
```

```
550 IF R$="Y" OR R$="y" THEN 580
560 IF R$="N" OR R$="n" THEN 590
570 GOTO 530 'LOOP FOR CHOICES IN UPPER OR LOWER CASE
580 GOSUB 3750
590 CLS
600 PRINT "TYPE 'S' FOR SCREEN PRINT, OR TYPE 'P' FOR PRINTER"
610 INPUT Q$
620 IF Q$="S" OR Q$="s" THEN 640
630 IF Q$="P" OR Q$="p" THEN 700
640 IF B$="Y" OR B$="y" THEN 2170
650 IF Q$= "P" OR Q$="p" THEN 700
660 IF R$= "y" OR R$="Y" THEN 2150
670 IF R$= "N" OR R$="n" THEN 2270
680 IF B$="N" OR B$="n" THEN 2270
690 GOTO 610 'LOOP FOR CHOICES IN UPPER OR LOWER CASE
700 PRINT "TURN ON PRINTER AND POSITION PAPER, THEN TYPE 'P' FOR PRINT"
710 INPUT G$
720 IF G$="P" OR G$="p" THEN 740
730 GOTO 710 'LOOP FOR CHOICES IN UPPER OR LOWER CASE
740 IF R$="Y" OR R$="y" THEN 800
750 IF B$="y" OR B$="Y" THEN 780
760 IF B$="N" OR B$="n" THEN 900
770 IF R$="N" OR R$="n" THEN 900
```

```
780 LPRINT "CALCULATED CONSTANTS USING THE INTERNAL SUBROUTINE"
790 GOTO 810
800 LPRINT "INPUT OF YOUR VERSION OF THE FUNDAMENTAL PHYSICAL
CONSTANTS"
810 LPRINT
820 LPRINT A#"JOULE PER TESLA, MAGNETIC MOMENT"
830 LPRINT D#"AMPERE SECONDS, CHARGE OF THE ELECTRON"
840 LPRINT E#"METERS, RATIONALIZED COMPTON WAVELENGTH-ELECTRON"
850 LPRINT F#"KILOGRAMS, MASS OF THE ELECTRON"
860 LPRINT H1#"JOULE SECONDS, RATIONALIZED PLANCK CONSTANT"
870 IF R$="Y" OR R$="y" THEN 1030
880 LPRINT R1#"RYDBERG CONSTANT, PER METER"
890 GOTO 1030
900 LPRINT "EXACT 1986 CODATA FUNDAMENTAL PHYSICAL CONSTANTS"
910 LPRINT
920 A#=9.2847701D-24
930 LPRINT "MAGNETIC MOMENT OF ELECTRON=";A#;"JOULES PER TESLA"
940 D#=1.60217733D-19
950 LPRINT "CHARGE OF ELECTRON=";D#;"AMPERE SECONDS"
960 E#=3.86159323D-13
970 LPRINT "RATIONALIZED COMPTON WAVELENGTH ELECTRON=";E#;"METERS"
980 F#=9.10938972D-31
990 LPRINT "MASS OF ELECTRON=";F#;"KILOGRAMS"
```

```
1000 H1#= 1.05457266D-34
1010 LPRINT "RATIONALIZED PLANCK CONSTANT=";H1#;"JOULES  SECONDS"
1020 LPRINT "RYDBERG CONSTANT= 10973731.534  PER METER"
1030 DIM V#(19),S#(19)
1040 B#=AU#
1050 X#=SQR(2#)
1060 H#=X#
1070 J#=H#*B#
1080 K#=(1#-(H#/2#))/(2#)-J#/2#
1090 M#=(D#*D#)/(F#*12.566370614359 17#*C#*C#*E#*K#*G#)
1100 L#=1#-J#
1110 D2#=M#/L#
1120 P#=1#/L#
1130 S#(1)=M#
1140 V#(1)=1#
1150 N#=H#/2#
1160 FOR R=1 TO 18
1170 D1#=N#^R
1180 V#(R+1)=P#/D1#
1190 S#(R+1)=D2#/D1#
1200 NEXT R
1210 LPRINT
1220 LPRINT
```

```
1230 LPRINT
1240 IF R$="Y" OR R$="y" THEN 1260
1250 IF R$="N" OR R$="n" THEN 1280
1260 LPRINT "   USING THE ABOVE ADJUSTED PHYSICAL CONSTANTS GIVES THIS"
1270 GOTO 1290
1280 LPRINT "   USING THE ABOVE 1986 CODATA PHYSICAL CONSTANTS GIVES THIS"
1290 LPRINT "      INTERNAL STRUCTURE FOR THE PROTON AND NEUTRON"
1300 LPRINT  "      ******************************************"
1310 LPRINT "SITE     PHOTON EFF    ELECTRIC POTENTIAL     TOTAL"
1320 LPRINT "  N     MASS RATIO    ENERGY MASS RATIO   MASS RATIO"
1330 LPRINT "----    ------------    ----------------   ---------"
1340 T#=0
1350 T1#=0
1360 T2#=0
1370 FOR Z=1 TO 19 STEP 2
1380 T1#=T1#+V#(Z)
1390 T2#=T2#+S#(Z)
1400 T#=T#+V#(Z)+S#(Z)
1410 D3#=V#(Z)+S#(Z)
1420 G$="ELECTRON MASSES"
1430 D$="  ##   ###.#########   + ###.#########    =   ###.#########"
1440 LPRINT  USING D$;Z;V#(Z);S#(Z);D3#
1450 IF Z=19 THEN 1510
```

```
1460 T1#=T1#+V#(Z+1)
1470 T2#=T2#+S#(Z+1)
1480 T#=T#+V#(Z+1)+S#(Z+1)
1490 D4#=V#(Z+1)+S#(Z+1)
1500 LPRINT  USING D$;Z+1;V#(Z+1);S#(Z+1);D4#
1510 NEXT Z
1520 W#=(T2#*((F#*C#*C#)/D#))*.000001
1530 LPRINT  "      -------------     ------------      -----------"
1540 E$="TOTAL ####.########   +  ##.########    =  ####.########  PROTON MASS"
1550 LPRINT  USING E$;T1#;T2#;T#
1560 LPRINT "ELECTRICAL POTENTIAL ENERGY OF"W#"MEV HOLDS THE PROTON
TOGETHER"
1570 LPRINT
1580 R#=T#+V#(1)+V#(2)+S#(1)+S#(2)
1590 IF T#>=1836.152701#  THEN 1620
1600 C1#=((1836.152701#/T#)-1)/.000001#
1610 GOTO 1630
1620 C1#=((T#/1836.152701#)-1)/.000001#
1630 IF R#>=1838.68366#  THEN 1660
1640 C2#=((1838.68366#/R#)-1)/.000001#
1650 GOTO 1670
1660 C2#=((R#/1838.68366#)-1)/.000001#
1670 LPRINT "1986 CODATA PROTON MASS RATIO = 1836.152701"
```

```
1680 F$="     MODEL  PROTON MASS RATIO = ####.###########"
1690 LPRINT  USING F$;T#
1700 LPRINT"       (DELTA";C1#;"ppm)"
1710 LPRINT
1720 LPRINT"                          1986 CODATA NEUTRON MASS RATIO = 1838.68366"
1730 L$="PAIRS AT SITE 1 AND 2 GIVE A NEUTRON MASS RATIO = ####.##########"
1740 LPRINT  USING L$;R#
1750 LPRINT "            (DELTA";C2#;"ppm)"
1760 N#=H#/2#
1770 A1#=N#^18
1780 N#=H#/2#
1790 U1#=(.5#*C#*D#*E#)*(A1#*L#)*(1#+(B#/2#))*(1#+(B#/2#))
1800 LPRINT
1810 J$="J/T"
1820 LPRINT  "PROTON'S TRUE MAGNETIC MOMENT =";U1#;"J/T"
1830 LPRINT
1840 A2#=E#*(A1#*L#)*((1#-(H#/2#))/2#)
1850 A3#=(J#/2#)*E#*(A1#*L#)
1860 A4# =A2#-A3#
1870 A5#=((1#-(H#/2#))/2#)-B#
1880 A6#=((H#/2#)+(B#/2#))
1890 B1#=A5#/A6#
1900 P1#=1#/(U1#-(U1#*B1#))
```

```
1910 M1#=D#/(H1#*C#*C#)
1920 M2#=M1#/(12.56637061435917#*G#*A4#)
1930 U2#=1#/(P1#+M2#)
1940 LPRINT"              1986 CODATA PROTON MAGNETIC MOMENT= 1.41060761E-26 J/T"
1950 LPRINT   "PROTON'S GYROSCOPICALLY CORRECTED MAGNETIC MOMENT
UG=";U2#;"J/T"
1960 IF U2#>=1.41060761D-26  THEN 1990
1970 C3#=((1.41060761D-26/U2#)-1)/.000001#
1980 GOTO 2000
1990 C3#=((U2#/1.41060761D-26)-1)/.000001#
2000 LPRINT"                        (DELTA";C3#;"ppm)"
2010 IF (C1#<3.3  AND C1#>3 ) THEN 2050
2020 PRINT
2030 LPRINT"YOU DID NOT USE A CONSISTENT SET OF CONSTANTS, TRY AGAIN!!!!"
2040 GOTO 2050
2050 IF (U2#<1.41067D-26  AND U2#>1.4106D-26)  THEN 2090 ELSE 2060
2060 LPRINT
2070 LPRINT"NOTE:  WITH FICTITIOUS SET OF CONSTANTS, PROTON "
2080 LPRINT"     MAGNETIC MOMENT PREDICTION IS WRONG VALUED"
2090 GOTO 3480
2100 LPRINT
2110 GOTO 3510
2120 CLS
```

```
2130 IF B$="Y" OR B$="y" THEN 2170
2140 IF B$="n" OR B$="N" THEN 2270
2150 PRINT"INPUT OF YOUR VERSION OF THE CONSTANTS"
2160 GOTO 2180
2170 PRINT"CALCULATED CONSTANTS FROM INTERNAL SUBROUTINE"
2180 PRINT
2190 PRINT A#"JOULE PER TESLA, MAGNETIC MOMENT OF ELECTRON"
2200 PRINT D#"AMPERE SECONDS, CHARGE OF ELECTRON"
2210 PRINT E#"METERS, RATIONALIZED COMPTON WAVELENGTH-ELECTRON"
2220 PRINT F#"KILOGRAMS, MASS OF ELECTRON"
2230 PRINT H1#"JOULE SECONDS, RATIONALIZED PLANCK CONSTANT"
2240 IF R$="Y" OR R$="y" THEN 2430
2250 PRINT R1#"RYDBERG CONSTANT, PER METER"
2260 GOTO 2430
2270 PRINT"EXACT 1986 CODATA FUNDAMENTAL PHYSICAL CONSTANTS"
2280 PRINT
2290 A#=9.2847701D-24
2300 PRINT "MAGNETIC MOMENT OF ELECTRON=";A#;"JOULES FER TESLA"
2310 C#=299792458#
2320 PRINT "VELOCITY OF LIGHT=";C#;"METERS PER SECOND"
2330 D#=1.60217733D-19
2340 PRINT "CHARGE OF ELECTRON=";D#;"AMPERE SECONDS"
2350 E#=3.86159323D-13
```

```
2360 PRINT "RATIONALIZED  COMPTON WAVELENGTH  ELECTRON=";E#;"METERS"
2370 F#=9.10938972D-31
2380 PRINT "MASS OF ELECTRON=";F#;"KILOGRAMS"
2390 PRINT "PERMITTIVITY OF SPACE=";G#;"FARAD  PER METER"
2400 H1#= 1.05457266D-34
2410 PRINT "RATIONALIZED PLANCK CONSTANT=";H1#;"JOULES  SECONDS"
2420 PRINT"RYDBERG CONSTANT=10973731.534  PER METER"
2430 DIM V#(19),S#(19)
2440 B#=AU#
2450 X#=SQR(2#)
2460 H#=X#
2470 J#=H#*B#
2480 K#=(1#-(H#/2#))/(2#)-J#/2#
2490 M#=(D#*D#)/(F#*12.56637061435917#*C#*C#*E#*K#*G#)
2500 L#=1#-J#
2510 D2#=M#/L#
2520 P#=1/L#
2530 S#(1)=M#
2540 V#(1)=1#
2550 N#=H#/2#
2560 FOR R=1 TO 18
2570 D1#=N#^R
2580 V#(R+1)=P#/D1#
```

```
2590 S#(R+1)=D2#/D1#
2600 NEXT R
2610 PRINT
2620 PRINT
2630 PRINT
2640 IF B$="Y" OR B$="y" AND R$="Y" OR R$="y" THEN 2660
2650 IF B$="N" OR B$="n" THEN 2680
2660 PRINT "   USING THE ABOVE ADJUSTED PHYSICAL CONSTANTS GIVES THIS"
2670 GOTO 2690
2680 PRINT "   USING THE ABOVE 1986 CODATA PHYSICAL CONSTANTS GIVES THIS"
2690 PRINT "      INTERNAL STRUCTURE FOR THE PROTON AND NEUTRON"
2700 PRINT  "      ********************************************"
2710 PRINT "SITE     PHOTON EFF    ELECTRIC POTENTIAL     TOTAL"
2720 PRINT "  N      MASS RATIO     ENERGY MASS RATIO   MASS RATIO"
2730 PRINT "----     ------------    ----------------   ----------"
2740 T#=0
2750 T1#=0
2760 T2#=0
2770 FOR Z=1 TO 19 STEP 2
2780 T1#=T1#+V#(Z)
2790 T2#=T2#+S#(Z)
2800 T#=T#+V#(Z)+S#(Z)
2810 D3#=V#(Z)+S#(Z)
```

```
2820 G$="ELECTRON MASSES"
2830 D$=" ##   ###.#########   + ###.#########    =    ###.#########"
2840 PRINT  USING D$;Z;V#(Z);S#(Z);D3#
2850 IF Z=19 THEN 2910
2860 T1#=T1#+V#(Z+1)
2870 T2#=T2#+S#(Z+1)
2880 T#=T#+V#(Z+1)+S#(Z+1)
2890 D4#=V#(Z+1)+S#(Z+1)
2900 PRINT  USING D$;Z+1;V#(Z+1);S#(Z+1);D4#
2910 NEXT Z
2920 W#=(T2#*((F#*C#*C#)/D#))*.000001
2930 PRINT  "      -------------      -----------      ----------"
2940 E$="TOTAL ####.########   +  ##.########    =  ####.########  PROTON MASS"
2950 PRINT  USING E$;T1#;T2#;T#
2960 PRINT "ELECTRICAL POTENTIAL ENERGY OF"W#"MEV HOLDS THE PROTON
TOGETHER"
2970 PRINT
2980 R#=T#+V#(1)+V#(2)+S#(1)+S#(2)
2990 IF T#>=1836.152701# THEN 3020
3000 C1#=((1836.152701#/T#)-1)/.000001#
3010 GOTO 3030
3020 C1#=((T#/1836.152701#)-1)/.000001#
3030 IF R#>=1838.68366# THEN 3060
```

```
3040 C2#=((1838.68366#/R#)-1)/.000001#
3050 GOTO 3070
3060 C2#=((R#/1838.68366#)-1)/.000001#
3070 PRINT"1986 CODATA PROTON MASS RATIO = 1836.152701"
3080 F$="   MODEL   PROTON MASS RATIO = ####.###########"
3090 PRINT  USING F$;T#
3100 PRINT"             (DELTA";C1#;"ppm)"
3110 PRINT
3120 PRINT"                         1986 CODATA NEUTRON MASS RATIO = 1838.68366"
3130 L$="PAIRS AT SITE 1 AND 2 GIVE A NEUTRON MASS RATIO = ####.##########"
3140 PRINT  USING L$;R#
3150 PRINT"             (DELTA";C2#;"ppm)"
3160 N#=H#/2#
3170 A1#=N#^18
3180 N#=H#/2#
3190 U1#=(.5#*C#*D#*E#)*(A1#*L#)*(1#+(B#/2#))*(1#+(B#/2#))
3200 PRINT
3210 J$="J/T"
3220 PRINT  "PROTON'S TRUE MAGNETIC MOMENT =";U1#;"J/T"
3230 PRINT
3240 A2#=E#*(A1#*L#)*((1#-(H#/2#))/2#)
3250 A3#=(J#/2#)*E#*(A1#*L#)
3260 A4# =A2#-A3#
```

```
3270 A5#=((1#-(H#/2#))/2#)-B#
3280 A6#=((H#/2#)+(B#/2#))
3290 B1#=A5#/A6#
3300 P1#=1#/(U1#-(U1#*B1#))
3310 M1#=D#/(H1#*C#*C#)
3320 M2#=M1#/(12.56637061435917#*G#*A4#)
3330 U2#=1#/(P1#+M2#)
3340 PRINT"          1986 CODATA PROTON MAGNETIC MOMENT= 1.41060761E-26 J/T"
3350 PRINT   "PROTON'S GYROSCOPICALLY CORRECTED MAGNETIC MOMENT
UG=";U2#;"J/T"
3360 IF U2#>=1.41060761D-26  THEN 3390
3370 C3#=((1.41060761D-26/U2#)-1)/.000001#
3380 GOTO 3400
3390 C3#=((U2#/1.4106076D-26)-1)/.000001#
3400 PRINT"             (DELTA";C3#;"ppm)"
3410 IF (C1#<3.3  AND C1#>3) THEN 3440
3420 PRINT"YOU DID NOT USE A CONSISTENT SET OF CONSTANTS, TRY AGAIN!!!!"
3430 GOTO 3440
3440 IF (U2#<1.41067D-26  AND U2#>1.4106D-26)  THEN 3480 ELSE
3450 3450 PRINT
3460 PRINT"NOTE: WITH A FICTITIOUS SET OF CONSTANTS, PROTON "
3470 PRINT"     MAGNETIC MOMENT PREDICTION IS WRONG VALUED"
3480 STOP
```

```
3490 PRINT
3500 PRINT
3510 STOP
3520 CLS
3530  PRINT  "CHOOSE ANY MASS FOR THE ELECTRON, OR ENTER THE 1986 CODATA"
3540  PRINT  "VALUE, WHICH IS;      (9.10938972D-31)"
3550 INPUT M8#
3560 CLS
3570 L1#= M8#/KM#
3580 'CALC RYDBERG CONSTANT JUST FOR FUN'
3590 R1#=(A8#*A8#)/((2#*L1#)*(2#*3.141592653589793#))
3600 'CALC RATIONALIZED PLANCK CONSTANT H1'
3610 H2#=M8#*C#*L1#
3620 'CALC ELECTRON CHARGE E1'
3630 H#=H2#*(2*PI#)
3640 E1#=SQR((2#*A8#*H#)/(U#*C#))
3650 'CALC ELECTRON MAGNETIC MOMENT'
3660 UB#=((L1#*E1#*C#)/2)
3670 AU#=2*((SQR(1+AE#))-1)
3680 UE#=UB#*((1+(AU#/2))^2)
3690 A#=UE#
3700 E#=L1#
3710 D#=E1#
```

```
3720 F#=M8#
3730 H1#=H2#
3740 RETURN
3750 'ENTER VARIABLES FOR THE CONSTANTS'
3760 PRINT"INPUT OF YOUR VERSION FOR THE CONSTANTS"
3770 PRINT
3780 PRINT"ELECTRON MAGNETIC MOMENT, JOULE PER TESLA?"
3790 PRINT"YOU MAY ENTER THE 1986 CODATA VALUE, WHICH IS; "
3800 PRINT"  9.2847701D-24  OR A CONSISTENT VALUE FOR THE ELECTRON'S"
3810 PRINT"MAGNETIC MOMENT THAT WAS PREVIOUSLY CALCULATED."
3820 INPUT A#
3830 CLS
3840 PRINT"ELECTRON CHARGE, AMPERE SECONDS?"
3850 PRINT"YOU MAY ENTER THE 1986 CODATA VALUE, WHICH IS;"
3860 PRINT"  1.60217733D-19  OR A CONSISTENT VALUE FOR THE ELECTRON'S"
3870 PRINT"CHARGE THAT WAS PREVIOUSLY CALCULATED"
3880 INPUT D#
3890 CLS
3900 PRINT"ELECTRON RATIONALIZED COMPTON WAVELENGTH, METERS?"
3910 PRINT"YOU MAY ENTER THE 1986 CODATA VALUE, WHICH IS;"
3920 PRINT" 3.86159323D-13  OR A CONSISTENT VALUE FOR THE ELECTRON'S"
3930 PRINT"RATIONALIZED COMPTON WAVELENGTH THAT WAS PREVIOUSLY
CALCULATED"
```

```
3940 H1#=H2#
3950 INPUT E#
3960 CLS
3970 PRINT"ELECTRON MASS, KILOGRAMS?"
3980 PRINT"YOU MAY ENTER THE 1986 CODATA VALUE, WHICH IS;"
3990 PRINT" 9.10938972D-31  OR A CONSISTENT VALUE FOR THE ELECTRON'S"
4000 PRINT"MASS THAT WAS PREVIOUSLY CALCULATED"
4010 INPUT F#
4020 CLS
4030 PRINT"RATIONALIZED PLANCK CONSTANT, JOULE SECONDS?"
4040 PRINT"YOU MAY ENTER THE 1986 CODATA VALUE, WHICH IS;"
4050 PRINT" 1.05457266D-34   OR A CONSISTENT VALUE FOR THE RATIONALIZED"
4060 PRINT"PLANCK CONSTANT THAT WAS PREVIOUSLY CALCULATED."
4070 INPUT H1#
4080 CLS
4090 RETURN
```